AF902254

HYDRO-GEOMORPHOLOGY - MODELS AND TRENDS

Edited by **Dericks P. Shukla**

Hydro-Geomorphology - Models and Trends
http://dx.doi.org/10.5772/65532
Edited by Dericks P. Shukla

Contributors

Bartłomiej Szypuła, Ricardo Castedo, Carlos Paredes, Rogelio De La Vega-Panizo, Anastasio P. Santos, Ali Assani, Konstantinos Kaffas, Vlassios Hrissanthou, Dericks Shukla

Published by InTech
Janeza Trdine 9, 51000 Rijeka, Croatia

© The Editor(s) and the Author(s) 2017
The moral rights of the editor(s) and the author(s) have been asserted.
All rights to the book as a whole are reserved by InTech. The book as a whole (compilation) cannot be reproduced, distributed or used for commercial or non-commercial purposes without InTech's written permission. Enquiries concerning the use of the book should be directed to InTech's rights and permissions department (permissions@intechopen.com).
Violations are liable to prosecution under the governing Copyright Law.

Individual chapters of this publication are distributed under the terms of the Creative Commons Attribution 3.0 Unported License which permits commercial use, distribution and reproduction of the individual chapters, provided the original author(s) and source publication are appropriately acknowledged. More details and guidelines concerning content reuse and adaptation can be found at http://www.intechopen.com/copyright-policy.html.

Notice

Statements and opinions expressed in the chapters are these of the individual contributors and not necessarily those of the editors or publisher. No responsibility is accepted for the accuracy of information contained in the published chapters. The publisher assumes no responsibility for any damage or injury to persons or property arising out of the use of any materials, instructions, methods or ideas contained in the book.

Publishing Process Manager Dajana Pemac
Technical Editor SPi Global
Cover InTech Design team

First published October, 2017
Printed in Croatia
Legal deposit, Croatia: National and University Library in Zagreb

Additional hard copies can be obtained from orders@intechopen.com

Hydro-Geomorphology - Models and Trends, Edited by Dericks P. Shukla
p. cm.
Print ISBN 978-953-51-3573-9
Online ISBN 978-953-51-3574-6

We are IntechOpen, the first native scientific publisher of Open Access books

3,350+
Open access books available

108,000+
International authors and editors

115M+
Downloads

151
Countries delivered to

Our authors are among the
Top 1%
most cited scientists

12.2%
Contributors from top 500 universities

WEB OF SCIENCE™

Selection of our books indexed in the Book Citation Index
in Web of Science™ Core Collection (BKCI)

Interested in publishing with us?
Contact book.department@intechopen.com

Numbers displayed above are based on latest data collected.
For more information visit www.intechopen.com

For EU product safety concerns:
IN TECH d.o.o., Prolaz Marije Krucifikse Kozulić 3, 51000 Rijeka, Croatia,
info@intechopen.com or visit our website at intechopen.com.

Meet the editor

Dr. Dericks P. Shukla, assistant professor in the School of Engineering, Indian Institute of Technology Mandi (IIT Mandi), has more than 7 years of experience in the areas of remote sensing and GIS, natural hazard assessment and mapping, climatic-tectono-geomorphology and hydro-geochemistry. He has published more than 17 research papers in peer-reviewed journals and various conference papers. He has been working in the field of landslide susceptibility/hazard zonation and prediction, groundwater pollution mainly arsenic contamination and fluvial and glacial geomorphologic studies using remote sensing and GIS techniques. He has five ongoing projects as principal investigator funded by DLR (German Aerospace Agency); Space Application Centre, Indian Space Research Organization (SAC-ISRO); and Department of Science and Technology (DST).

Contents

Preface

"There is nothing permanent except change." — *Heraclitus*

Change is the law of life, and everything surrounding us has changed and will change in the future. However the timescale varies from discipline to discipline. The red blood cells in our body can be developed within 7 days from stem cells, while it takes 7 years for giant Himalayan lily—or *Cardiocrinum giganteum* var. yunnanense—to flower once, while in the case of earth science, most of the events that occur have a life cycle of thousands to millions of years. Hence to comprehend the forms and the processes that have led to formation of various earth surface features, understanding of geomorphology is necessary. For that the basic theme, "Present is the Key to the Past" is being followed, where the features that we see now are being studied and related to get the understanding of the processes of the past. For detailed understanding, many areas of geomorphology have been evolved such as fluvial geomorphology, glacial geomorphology, coastal, Aeolian tectonic, climatic geomorphology and joining them all is historical geomorphology. For quantitative estimation, another new field of computational geomorphology has surfaced recently with the advent of computer science.

Keeping these views in mind, this book on *Geomorphology* was envisaged. However, to compile all the aspects of geomorphology in just one book will be Herculean task; hence the focus of this book lies in the water-created features. Thus fluvial and coastal geomorphology has been focused in this book with a chapter on application of digital elevation model (DEM) in geomorphology. This book has been prepared with four groups of authors and their contributions. In the first chapter, just a basic idea and understanding of geomorphology has been presented, followed by Chapter 2 by **Kaffas and Hrissanthou**, who present a composite mathematical modelling for continuous simulations of hydro-geomorphological processes, as well as continuous simulations of soil and streambed erosion processes. In Chapter 3, **Assani et al.** study the impacts of deforestation and then reforestation along with the hydro-climatic variability on the morphology (width and sinuosity) of the Matambin River channel in Quebec, Canada. Both Chapters 2 and 3 present the case of fluvial geomorphology, while modelling of cliff erosion, a part of coastal geomorphology, is presented in Chapter 4. **Castedo et al.** present process-response models for estimation of cliff erosion and its quantitative predictions due to the effects of natural and human-induced changes. Finally in Chapter 5, **Szypuła** presents a detailed review by explaining various geomorphometric indices and parameters and shows the utilization of DEM for extraction of these information. He explains these tools with various examples that are available in various GIS packages.

Dericks P. Shukla
IIT Mandi
Mandi, India

Section 1

Geomorphology

Chapter 1

Introductory Chapter: Geomorphology

Dericks Praise Shukla

Additional information is available at the end of the chapter

http://dx.doi.org/10.5772/intechopen.70823

1. Introduction

Over millions of years, the Earth has gone through many changes which have shaped its current form and structure. From a dust ball according to nebular hypothesis, to the current form, the Earth has transformed a lot. Once an inhabitable place, during the Hadean time, our Earth has seen many processes over a long time of more than 4 billion years. Developmental stages which formed the current habitable world include both internal and external forces. The meteoritic impact, volcanic activities, and erosional activities of rivers, winds, glaciers, oceans, etc. along with the sea floor spreading and plate tectonic activities have been constantly working to shape the Earth as we see now. Many of these activities occur during a short interval, while some take millions of years to create various climatic, geologic, and geomorphic regimes. All of these never-ending processes are still continuously going on and shaping our Earth currently. The most notable of all these processes are geomorphic processes since they create the shape and form of the Earth as we see it now. Hence, the study of these geomorphic processes is critical to understand the phenomena and process that are occurring in nature.

Having its derivation from Greek words, γεω (Earth), μορφη (morph/form), and λογος (discuss), geomorphology literally means "a discussion on Earth's form." Hence, it is the study of various features that are found on the Earth, such as mountains, hills, plains, rivers, moraines, cirques, sand dunes, beaches, spits, etc., that are created by various agents such as rivers, glaciers, wind, ocean, etc. Since the fourth century BC, many people have studied the formation of the Earth by relating to various observations in the field. Ancient Greeks and Romans such as Aristotle, Strabo, Herodotus, Xenophanes, and many others discussed about the origin of the valleys, formation of deltas, presence of seashells on mountains, etc. After observing the seashells on the top of the mountains, Xenophanes speculated that the surface of the Earth must have risen and fallen from time to time, thus creating river valleys and mountains (*c.* 580–480 BC). After observing seashells on mountain top and vast expanses of sand, Aristotle (*c.* 384–322 BC) suggested that the areas which are dryland now must be

INTECH
open science | open minds

© 2017 The Author(s). Licensee InTech. This chapter is distributed under the terms of the Creative Commons Attribution License (http://creativecommons.org/licenses/by/3.0), which permits unrestricted use, distribution, and reproduction in any medium, provided the original work is properly cited. (cc) BY

covered by sea in the past and those areas where sea is present now must have been dryland once. Hence, he proposed that land and sea change places. Traditionally, the history of the development of landscape was carried out by mapping the sedimentary and morphological features. For understanding the evolution of landscape, the golden rule, "the present is the key to the past," has been followed. This rule assumes that the processes that are visible in action today must have occurred in the past also, which can be used to infer the reasons for formation of the landscape in the past. Hence, the past formation was mainly dependent on the relative information and aging method.

However, the word "geomorphology" was first coined and used between the 1870s and 1880s [1] to describe the morphology of the surface of the Earth. But it was popularized by William Morris Davis who proposed the "geographical cycle" also known as "Davis cycle" [2]. He proposed that the development of landscapes occurs as due to alternate action of uplift and denudation. He assumed that uplift occurs quickly and then the uplifted land mass erodes gradually to form the topography of the region. He hypothesized that upliftment is a quick action, whereas denudation is a time-taking process. Thus, creating high mountains and deep valleys showcases youth, mature, and old age stages of landform development. Though Davis cycle is considered as a classic work, but his hypothesis lacks a basic understanding that both upliftment and denudation occur simultaneously. Both of these phenomena go on hand in hand and are not necessarily alternate. Hence, nearly 35 years later, Walther Penck proposed a variation of "Davis model," where he showed that the interaction of both uplift and denudation occurs simultaneously [3, 4]. He suggested that due to simultaneous actions, the slopes will be developed in three main forms. First, a convex slope where the upliftment rate is higher than denudation rate; next, a steady-state or stationary stage where both the rates are nearly equal, hence creating straight slope; and finally concave slopes when the rate of upliftment is lesser than the rate of denudation. Thus, over a period of time, various aspects of landforms have been studied by geomorphologists. Some geomorphologists have studied the process of formation of these landforms, while some have studied its origin and history, and others have analyzed various forms of landforms for their quantitativeness. Hence, in a nutshell, modern geomorphologists focus mainly on three aspects of landforms: form, process, and history. The form and process studies are commonly termed as functional geomorphology, while the last one as historical geomorphology [5]. The study of various processes that are responsible for the formation of a landscape falls in the purview of functional geomorphology.

All these landforms that are visible on the Earth vary in size from microscale such as potholes, flutes, ripples, etc. to mega-scale features such as mountain ranges, river basins, etc. Hence, the time required to form these features also varies from tens of years to millions of years. It has also been observed that certain features are native to certain climatic zones; hence, development of climatic zones such as arid, tropical, etc. plays a critical role in formation and evolution of these geomorphic features. For example, the landforms observed in higher latitudes show signature of glaciation and deglaciation cycle, which is indicative of quaternary climatic environment, whereas in other parts of the world, such as Grand Canyon of Colorado River Valley in the United States have preserved the signature of various activities, that occurred hundreds of millions of years ago, in its various landforms. Most of the landforms are formed and deformed due to the two processes, namely, endogenic that occurs within the Earth's crust such as convective heat cycles, rising plume, and magma chambers and exogenic processes that shape the features on the surface of the Earth with the help of various agents of weathering like water, wind, glaciers, seas, etc. All these phenomena of landscape evolution with respect to life span, climatic zones, and processes are shown in **Figure 1**.

A lot of works have been carried out in the field of functional and historical geomorphology. Now, many other fields or kinds of geomorphology have been studied such as tectonic geomorphology, submarine geomorphology, planetary geomorphology, climatic geomorphology, and modeling geomorphology. Interaction of tectonic forces and geomorphic processes deform the Earth's crust regularly, and this led to development of tectonic geomorphology. It uses the techniques and data from other fields of geology mainly structural, geochemistry, geochronology in conjunction with geomorphology, and climate change. As name suggests, submarine geomorphology focuses on the origin, formation, and development of submarine landforms developed in both shallow and deep marine environments. Planetary geomorphology deals with the application of the understanding of the formation of landforms on the Earth to extraterrestrial objects such as moon, planets, exoplanets, etc. This is comparatively the latest branch and is developing very fast. Geomorphic studies of Venus, Mars, Jupiter, Titan, and other planets are a hot cake these days. Climate plays a critical role in developing various landform natives to each climatic zone such as arid, tropical, temperate, etc. This understanding is the basis for the development of climatic geomorphology as a stream. The effect of climatic phenomena along with tectonic activities places a new cross

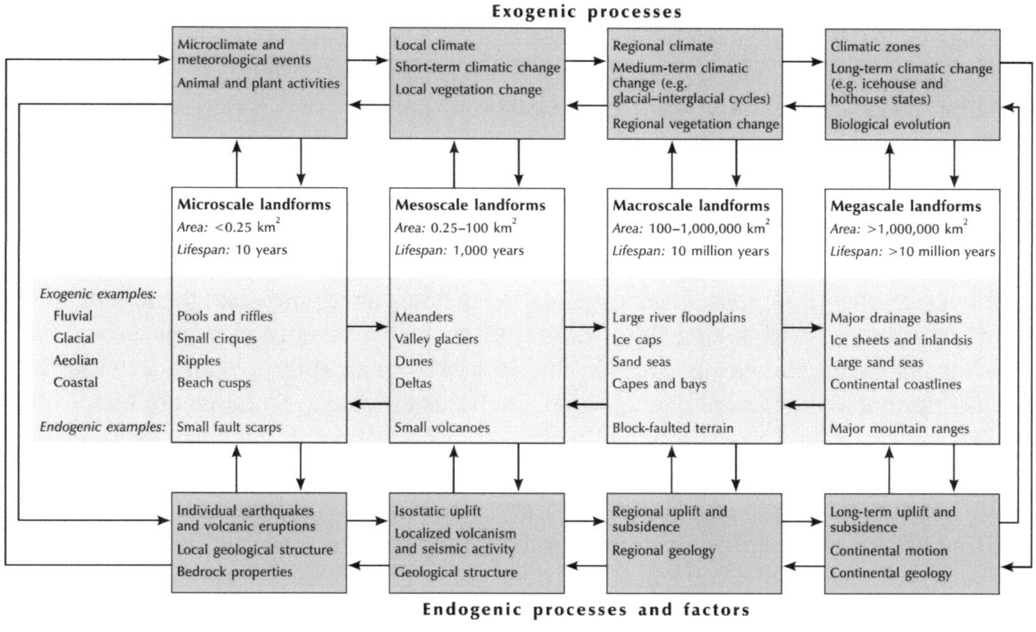

Figure 1. Form, process, and their interrelationships for the evolution of various landforms developed due to endogenic and exogenic processes over various scales of time and climatic zones (adapted from [6]).

stream of geomorphology known as climato-tectonic geomorphology. These days, inter- and multidisciplinary approaches have been used in various fields of science, and geomorphology is one of them where cross breeding is highly evident. Till now, various branches and offshoots of geomorphology have been developed, and lots of researches have been carried out in those interdisciplinary areas.

Among all the exogenic agents that are at work to form the landscape, water is the most promising and effective. Hence, fluvial geomorphology has been studied a lot in details. Keeping these aspects in mind, this book has been formulated where the major focus is on geomorphic features developed due to action of water. Hence, two chapters on fluvial geomorphology and one chapter on coastal geomorphology are being presented here. While the last chapter deals with the recent trends of digital elevation model (DEM) that could be very effectively used for morphometric analysis of various streams.

Hydro-geomorphology, the study of hydrological processes, involves surface runoff, baseflow, stream discharge, and the soil and streambed erosion processes, which continuously chisel the geomorphological profile of a basin. The life span of such processes varies from few hundreds of years to even millions of years. Apart from the quantification of the hydrological processes, as well as the soil and streambed erosion processes, the continuous hydro-geomorphologic modeling provides valuable information for the future trend of these physical processes. A wide variety of integrated models that continuously simulate the runoff, soil erosion, and sediment transport processes are available. In Chapter 2, a composite mathematical model is presented aiming at continuous simulations of hydro-geomorphological processes, as well as continuous simulations of soil and streambed erosion processes in Kosynthos River basin (district of Xanthi, Thrace, northeastern Greece) and Nestos River basin (Macedonia-Thrace border, northeastern Greece), the two neighboring basins in northeastern Greece. Their model generates continuous hydrographs and sediment graphs at the outlets of the two basins at fine temporal scales, whose statistical efficiency with the measured quantities at the basin outlet is highly significant providing satisfactory results. The correlation coefficient of modeled values with measured values is more than 80% for both the basins for water and sediment discharge.

Anthropogenic activities have significantly affected the fluvial geomorphological regimes within a very short time span. From construction of dams which increases the sedimentation in the reservoir, thus changing the riverbed profile to the deforestation and urbanization which is enhancing the erosion rates in the river catchment, anthropogenic activities have left its imprints in the natural phenomenon. Similar is the case in St. Lawrence Lowlands of Quebec region of the Canadian Shield where the construction of dams has led to increase in bank-full width, thus decreasing the channel sinuosity and changing the fluvial regimes. Further changes in land-use pattern have also led to higher erosion and sedimentation. Clearing of forests for agricultural practices has led to deforestation, and later afforestation in that region (agricultural areas) due to decline in agricultural work force has impacted the morphological evolution of channels in Quebec region of Canada. Thus, Chapter 3 tries to constrain the impacts of reforestation and hydroclimatic variability on the morphology (width and sinuosity) of the Matambin River channel in Quebec, Canada. They observed

21% decrease in mean channel bank-full width from 1935 to 1964 which was characterized by a low frequency of strong flood flows in the region. After 1964, a trend of increasing mean channel bank-full width was observed which is associated with the increase in frequency of strong flood flows and decrease in the amount of suspended sediments produced by soil erosion.

Higher rates of erosion are observed when the weathering agent is water. And, considering the huge expanses of oceans and the erosion that occurs at the shores takes the first place. This effect is clearly visible in shoreline change and sea-level rise. Most of the populated cities all around the world are situated near the coasts; thus, majority of the population of the world lives within few kilometers of the coast. Thus, a proper coastal land management is required to cater the needs of the ever-increasing population. Shoreline change (cliff erosion) has been studied using predictive models which are based on historical records and the geomorphological data of a certain region. Current historical extrapolation models use historical recession data, but different environments with the same historical values can produce identical annual retreat characteristics despite the potential responses to a changing environment being unequal. For that reason, process-response models are being explained in Chapter 4 based on real data at Holderness coast (UK), to provide quantitative predictions of the effects of natural and human-induced changes that cannot be predicted using other models.

With the advent of satellite technology, it has been absolutely easy to study the surface of the Earth from satellite data. When it comes to identifying various landforms and describing the physical appearances, satellite images or aerial photos come very handy. However, this approach is more qualitative than quantitative and is defined as morphography, where the external shapes are described without giving information about the way of creation of those features. Various methods are used to define the origin of features and the mechanism of development of these features. This comes under morphogenesis, while morphochronology deals with the estimation of age of the forms in the absolute as well as relative terms. Finally, the quantitative estimation carried out by measurements of the geometric features of the landforms is known as morphometry. There are various morphometric parameters and morphometric indices being used in geomorphometry to define the landform analysis and classification. Chapter 5 presents a detailed review by explaining various geomorphometric indices and parameters and shows the utilization of DEM for extraction of these information. He explains these tools with various examples that are available in various GIS packages.

Author details

Dericks Praise Shukla

Address all correspondence to: dericks.82@gmail.com

Indian Institute of Technology Mandi (IIT Mandi), India

References

[1] de Margerie E. Géologie. Polybiblion Revue Bibliographique Universelle, Partie littéraire. 1886;**2**(24):310-330

[2] Davis WM. The rivers and valleys of Pennsylvania. National Geographical Magazine. 1889;**1**:183-253 (Also in Geographical Essays)

[3] Penck W. Die morphologische Analyse, ein Kapitel der physikalischen Geologie. Stuttgart: Engelhorn; 1924

[4] Penck W. In: Czech H, Boswell KC, editors. Morphological Analysis of Landforms. London: Macmillan; 1953

[5] Chorley RJ. Bases for theory in geomorphology. In: Embleton C, Brunsden D, Jones DKC, editors. Geomorphology: Present Problems and Future Prospects. Oxford: Oxford University Press; 1978. pp. 1-13

[6] Huggett RJ. Fundamentals of Geomorphology. 2nd ed. Routledge Taylor & Francis Group, London and New York, 2007

Chapter 2

Computation of Hydro-geomorphologic Changes in Two Basins of Northeastern Greece

Konstantinos Kaffas and Vlassios Hrissanthou

Additional information is available at the end of the chapter

http://dx.doi.org/10.5772/intechopen.68655

Abstract

This chapter presents a composite mathematical model aiming at continuous simulations of hydro-geomorphological processes at the basin scale. Continuous hydrologic simulations, as well as continuous simulations of soil and streambed erosion processes, are performed in two neighbouring basins in northeastern Greece: Kosynthos river basin (district of Xanthi, Thrace, northeastern Greece) and Nestos river basin (Macedonia-Thrace border, northeastern Greece). Both basins are mountainous and covered by forested and bushy areas in their greatest part. Kosynthos river basin extends to an area of 237 km^2, whilst Nestos river basin is quite bigger, covering an area of approximately 840 km^2. The characteristic of Nestos river basin is the presence of a dam at its northwestern boundary, which largely affects the discharge, as well as the sediment transport in Nestos River. The application of the model results in continuous hydrographs and sediment graphs at the outlets of the two basins. Fine temporal scales are used, providing this way a continuous assessment of water and sediment discharge. The statistic efficiency criteria utilized for the comparison between computed and measured values of water and sediment discharge at the basin outlet provide satisfactory results. Therefore, it is concluded that the continuous hydro-geomorphologic modelling can be successfully applied to Kosynthos and Nestos river basins.

Keywords: hydro-geomorphology, continuous hydrographs, soil erosion, sediment transport, continuous sediment graphs, sediment yield

1. Introduction

Hydro-geomorphology is primarily a matter of water and sediment. This constitutes the study of hydrological processes, as well as of the soil and streambed erosion processes, imperative. Surface runoff, baseflow, stream discharge, soil erosion and sediment transport constitute the

INTECH
open science | open minds

© 2017 The Author(s). Licensee InTech. This chapter is distributed under the terms of the Creative Commons Attribution License (http://creativecommons.org/licenses/by/3.0), which permits unrestricted use, distribution, and reproduction in any medium, provided the original work is properly cited. (cc) BY

basic interrelated natural processes, which perpetually chisel the geomorphological profile of a basin. In most cases, it takes hundreds or even thousands of years for any effect to take its toll on the geomorphological profile of the earth. However, it is worth studying the hydro-geomorphological processes at a continuous timescale. Apart from the quantification of the hydrological processes, as well as of the soil and streambed erosion processes, the continuous hydro-geomorphologic modelling provides valuable information for the future trend of these physical processes.

There is a wide variety of integrated models that continuously simulate the runoff, soil erosion and sediment transport processes. Just a few of these models are listed below. Most of these models are available in their software form:

1.1. Areal non-point source watershed environment response simulation model (ANSWERS)

ANSWERS [1] is a physically based, distributed model. Although the model was initially developed to operate on an event basis, it underwent modifications to be used for continuous simulations as well [2]. The current version of ANSWERS can continuously simulate runoff by means of the Holtan [3] or the Green and Ampt infiltration methods [4]. The continuity equation is the basic equation used for the computation of surface runoff:

$$\frac{dS}{dt} = I - O \tag{1}$$

where S is the volume of water in storage (surface detention), t is the time, I is the inflow rate and O is the outflow rate.

Soil erosion is modelled by estimating raindrop detachment using rainfall intensity and universal soil loss equation (USLE) factors [5]. As far as sediment transport is concerned, Yalin's bed load transport equation is used [6]:

$$T = 146 \cdot s \cdot Q^{\frac{1}{2}} \quad \text{for } Q \le 0.046 \, [\, m^3/(\min m)\,] \tag{2}$$

$$T = 14,600 \cdot s \cdot Q^2 \quad \text{for } Q \ge 0.046 \, [\, m^3/(\min m)\,] \tag{3}$$

where T is the transport capacity by surface runoff [kg/(min m)], s is the surface slope and Q is the flow rate per unit width [m^3/(min m)].

Yalin's equation was originally conceived for the routing of sediments through a channel. However, several attempts have been made towards the application of this channel formula for overland flow. Amongst these, Foster and Meyer [7], as well as Alonso et al. [8], proved that Yalin's equation can successfully be applied for sediment transport by surface runoff.

1.2. Agricultural non-point source model (AGNPS)

AGNPS [9, 10] is a conceptual distributed model, which operates on a cell basis. AGNPS was also initially developed for event-based simulations. Latest versions of the model, though, offer the ability for long-term hydrologic and soil erosion simulations. The modelling of runoff is based on soil conservation service (SCS)-curve number (CN) method [11] (Section 4.1).

Soil erosion and sediment transport on the soil surface are modelled using the revised universal soil loss equation (RUSLE) [12]:

$$A = R \cdot K \cdot L \cdot S \cdot C \cdot P \tag{4}$$

where A is the average soil loss per unit area and time [t/(ha yr)], R is the rainfall erosivity factor [MJ mm/(ha h yr)], K [t h/(MJ mm)] is the soil erodibility factor, L is the slope length factor, S is the slope steepness factor, C is the cover and management factor and P is the support practice factor.

The RUSLE is exactly the same with USLE. The only difference between RUSLE and USLE is that the factors K, L, S, C and P, in RUSLE, are computed in more detail.

Sediment routing in streams is modelled by a modified Einstein deposition equation [13] and the Bagnold suspended sediment formula for stream sediment transport [14].

1.3. Chemical runoff and erosion from agricultural management systems model (CREAMS)

Another physically based model which can continuously simulate the hydro-geomorphological processes, at the basin scale, is the chemical runoff and erosion from agricultural management systems model (CREAMS) [15]. As in the case of AGNPS, runoff modelling in CREAMS is based on the SCS-CN method for the estimation of hydrologic losses due to infiltration and kinematic surface water flow equations. The peak runoff rate is computed by the following equation:

$$q_p = 200\,(DA)^{0.7} \cdot (CS)^{0.15} \cdot Q^{0.917(DA)^{0.0166}} \cdot (LW)^{-0.187} \tag{5}$$

where q_p is the peak runoff rate (m^3/s), DA is the drainage area (km^2), CS is the channel slope, Q is the daily runoff volume (mm) and LW is the length-width ratio of the basin.

The erosion component maintains elements of the USLE but includes sediment transport capacity for overland flow, which is estimated by the steady-state continuity equation:

$$\frac{d\,q_s}{dx} = D_L + D_F \tag{6}$$

where q_s is sediment discharge per unit width [kg/(s m)], x is the distance downslope (m), D_L is the lateral inflow of sediment [kg/(s m^2)] and D_F is the detachment or deposition of sediment by flow[kg/(s m^2)].

The composite mathematical model (CMM), applied in this study, comprises three submodels: a rainfall-runoff submodel, a submodel for the simulation of soil erosion and a sediment transport submodel for the routing of sediments in streams. The rainfall-runoff submodel that is used for the computation of the overland flow, as well as of the flow in the mainstreams of the sub-basins, is the conceptual, semi-distributed hydrologic model HEC-HMS 4.2. The soil erosion submodel, utilized for the estimation of soil erosion in a sub-basin, is based on the relationships of Poesen [16]. The estimate of sediment yield at the outlet of a sub-basin, and finally at the outlet of the whole basin, is achieved by means of the stream sediment transport model of Yang and Stall [17]. The CMM was applied to Kosynthos river basin and to Nestos river basin. The application of all three models results in continuous hydrographs and sediment graphs at the basin outlet. The computed stream discharge and sediment discharge values are compared with field measurements, and all models are evaluated as to their competence of simulating the hydro-geomorphological processes in a basin.

The geomorphology of the earth's surface is shaped by physical, chemical, biological as well as geological processes. This chapter focuses on the effect that the physical processes have on the geomorphological profile of a soil surface. The component of physical processes mainly refers to the hydro-geomorphological processes.

2. Description of the study area

2.1. Kosynthos river basin

Kosynthos River originates from Mount Erymanthos in the mountain chain of central Rodopi, flows in a southeastern course, passes through the city of Xanthi and empties in Vistonida Lake close to the ancient city of Anastasioupoli. Its overall length is approximately 55 km. The study area concerns the mountainous part of the basin and extends to 237 km^2 (**Figure 1**) from the Greek-Bulgarian border, to the North, to the city of Xanthi and to the South and from the mountainous region of Livaditis, to the East, to the mountainous region between Myki and Kentavros and to the West. The altitude varies between 72 m and 1700 m, the average land

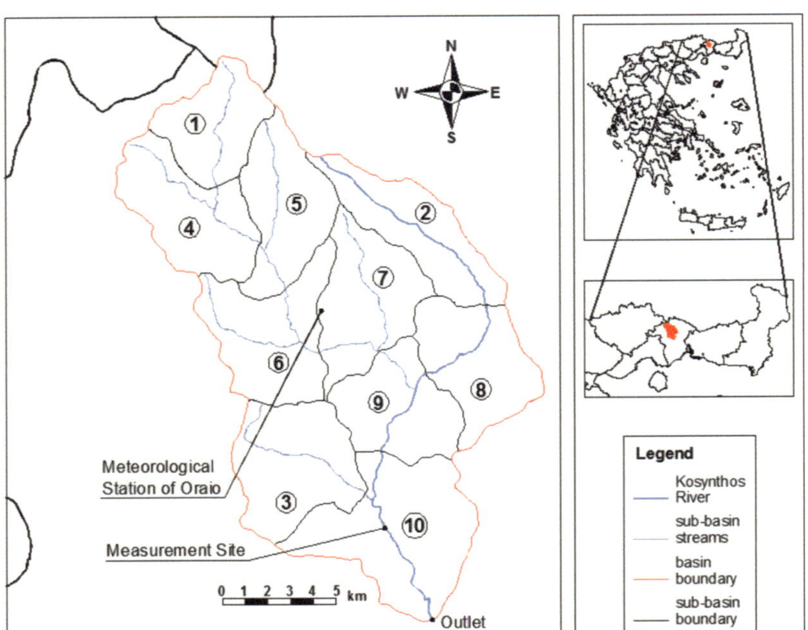

Figure 1. Kosynthos river basin.

slope of the basin is 37.3% and the length of the part of Kosynthos River that runs the basin is approximately 35 km.

The climate of the study area is of temperate Mediterranean type where the average annual temperature is 14°C and annual precipitation is about 750 mm.

The basin of Kosynthos River consists of forests (74%), bushes (4.5%), urban area (1.5%) and an area with no significant vegetation (sparse vegetation) (20%). The dominant rocks are granite-diorite, marble, gneiss-granite and migmatite. The structure of the semi-permeable soil in combination with the structure of the bedrock, which has a low percentage of deep percolation, favours a relatively high runoff discharge.

2.2. Nestos river basin

Nestos River straddles the border between Macedonia and Thrace in northeastern Greece (**Figure 2**). The study area of Nestos river basin falls in the Greek mountainous part downstream of Platanovrysi dam, which is located on the Macedonian-Thrace border, in northeastern Greece (**Figure 2**). The area of the basin is approximately 840 km^2. The altitude varies between 38 m and 1747 m, the average land slope of the basin is 37% and the length of the part of Nestos River that runs the basin is approximately 63 km.

The Platanovrysi dam (**Figure 2**) is operated by the Hydroelectric Power Production Agency of Hellenic Public Electricity Corporation. The daily dam discharges were provided by the

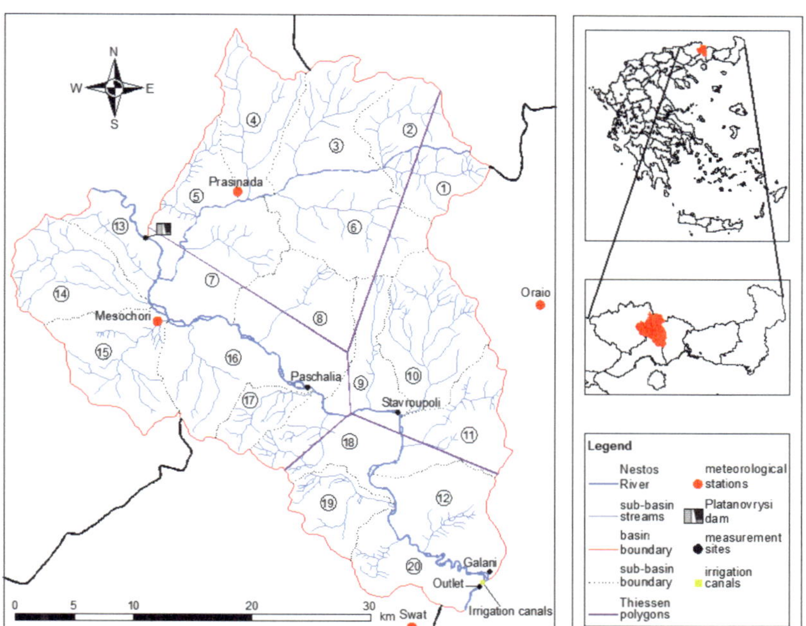

Figure 2. Nestos river basin.

Hydroelectric Power Production Agency for the entire study period (11 September 2005–31 July 2014). The two irrigation canals (**Figure 2**) are located at the left and right banks of the same cross-sectional area, very close to the Egnatia bridge of Nestos River, just slightly upstream of the basin outlet. The irrigation canals are operated by the local land reclamation agencies of Thalassia-Kremasti and Chrysoupoli. The irrigation season starts in the middle of April and goes up to the end of October; in some cases, it extends up to early November also.

The land cover data in combination with the soil permeability data are represented by the CN. The study area under investigation is covered mostly by forested and bushy areas and followed by crops, whilst a very small portion falls under urban areas and areas with no significant vegetation.

Soil texture data were obtained from soil associations' maps and were provided from Ref. [18]. There are four dominant soil types of the basin mainly: sandy clay loam (sand 55%, clay 19% and silt 26%), silty loam (sand 22%, clay 21% and silt 57%), loamy sand (sand 78%, clay 4% and silt 18%) and silty clay loam (sand 8%, clay 39% and silt 53%).

The bedrock mainly consists of semi-permeable rocks which do not favour deep percolation. The dominant rocks are marble, aluminous schist, rhyolite, lignite, schist, granite, granite-diorite, gneiss and gneiss-granite.

3. Meteorological data and field measurements

3.1. Kosynthos river basin

There is only one available meteorological station in the area of investigated part of Kosynthos river basin which has the data available from 1 January 2005 to 15 March 2009 (**Figure 1**). The meteorological data used for the application of the model are the following: rainfall depth (mm), minimum, maximum and average daily temperatures (°C), atmospheric pressure (kPa), wind speed at 2 m height (m/s) and solar radiation [MJ/(m^2 day)]. All of the aforementioned meteorological data were used at a daily time step for modelling. The meteorological station of Oraio is placed at the very centre of the basin (**Figure 1**), at an elevation of 800 m. Nonetheless, the rainfall data of only one meteorological station are neither sufficient nor representative of the whole basin, especially for the sub-basins close to the basin outlet.

The Laboratory of Hydrology and Hydraulic Structures, Civil Engineering Department, Democritus University of Thrace carried out 38 discharge measurements near the outlet of Kosynthos river basin, 5 of which in October, November and December 2005, 10 in March, April, May, June and July 2006, 14 in March, April and May 2007, 8 in May, June, July, September, November and December 2008 and 1 in January 2009. Additionally, 28 measurements of bed load and suspended load were carried out near the outlet of Kosynthos river basin, 4 of which in November, December 2005, 9 in March, April, May and June 2006, 6 in

April and May 2007, 8 in May, June, July, September, November and December 2008 and 1 in January 2009 (**Figure 1**).

Measurements of bed load transport and suspended load transport were carried out at the same time as the stream discharge measurements.

3.2. Nestos river basin

There are four meteorological stations in the Nestos river basin. Two of the stations are located inside the basin and the other two are little out of it (**Figure 2**). The data from the stations Mesochori and Prasinada, which lie inside the basin, were provided by the Hydroelectric Power Production Agency, the meteorological station of Oraio is operated by the Civil Engineering Department of Democritus University of Thrace, whilst the data from Soil and Water Assessment Tool (SWAT) station were obtained from the internet site Global Weather Data for SWAT [19] and were provided by the Ecosystem Sciences and Management Department of Texas A&M University, the USA. The meteorological data obtained from the meteorological stations are the rainfall depth (mm), minimum, maximum and average values of temperature (°C), wind speed (m/s), relative humidity and solar radiation (W/m^2). All these meteorological data are used at a daily time step. The distribution of the meteorological stations to areas of influence was achieved with Thiessen polygons [20] (**Figure 2**).

About 143 discharge measurements were carried out by the Laboratory of Hydrology and Hydraulic Structures of Civil Engineering Department, Democritus University of Thrace and the Laboratory of Ecological Engineering and Technology of Environmental Engineering Department, Democritus University of Thrace at four different intersections of the Nestos River—Paschalia, Stavroupoli, Galani and the basin outlet (**Figure 2**). These measurements were taken from September 2005 to May 2011, not on a regular basis. However, all months are covered through the above period. Moreover, the Laboratory of Hydrology and Hydraulic Structures carried out 40 measurements of bed load and suspended load at the outlet of the basin (**Figure 2**). Five of these measurements were carried out in September 2005 (**Figure 16**), whilst the rest from June 2008 to June 2011 (**Figure 17**).

4. Theoretical description of the composite mathematical model

4.1. Rainfall-runoff submodel

The purpose of the rainfall-runoff submodel is to simulate and quantify all the natural processes, taking place when rainfall starts, and to lead to the final transformation of rainfall-to-runoff hydrograph. The rainfall-runoff submodel consists of several components, each describing a natural process. These components are the rainfall excess, the evapotranspiration, lag time and time of concentration and the transformation of rainfall excess into runoff hydrograph. Since, in this case, we are interested in the simulation of the stream sediment transport processes as well, a flood routing model and a baseflow model are added. The basic equations of these methods are given below.

4.1.1. *Rainfall excess model*

The SCS-CN (NRCS since 1994) is used for the estimation of the hydrologic losses due to infiltration as well as for the estimation of rainfall excess [11]. The amount of rainfall excess is transformed into surface runoff.

$$Q = \frac{(P - 0.2S)^2}{P + 0.8S} \quad \text{for } P > 0.2S\text{, otherwise } Q = 0 \tag{7}$$

where Q is the rainfall excess (mm), P is the total rainfall (mm) and S is the maximum hydrologic losses (mm).

4.1.2. *Evapotranspiration*

For estimating the potential evapotranspiration, ET_o (mm/day), the widely known Penman-Monteith FAO-56 equation was used [21]:

$$ET_o = \frac{0.408\Delta(R_n - G) + \gamma \frac{900}{T + 273} u_2(e_s - e_a)}{\Delta + \gamma(1 + 0.34 u_2)} \tag{8}$$

where R_n is the net radiation at the crop surface [MJ/(m^2 day)], G is the soil heat flux density [MJ/(m^2 day)], T is the mean daily air temperature at height ranging from 1.5 m to 2.5 m (°C), u_2 is the wind speed at 2 m height (m/s), e_s is the saturation vapour pressure at height ranging from 1.5 m to 2.5 m (kPa), e_a is the actual vapour pressure at height ranging from 1.5 m to 2.5 m (kPa), Δ is the slope of the vapour pressure curve (kPa/°C) and γ is the psychrometric constant (kPa/°C).

4.1.3. *Lag time and time of concentration*

Lag time is calculated by means of the empirical equation of SCS [11]. SCS relates lag time, t_p, with the time of concentration, t_c:

$$t_p = 0.6t_c \tag{9}$$

Various equations are available for the estimation of the concentration time. In the present study, the most suitable was found to be Giandotti's formula [22, 23], for Kosynthos river basin, and Pasini's formula [24], for Nestos river basin.

4.1.4. *Transformation of rainfall excess into runoff hydrograph*

The transformation of rainfall excess into runoff hydrograph is achieved by means of the dimensionless synthetic unit hydrograph of soil conservation service [25].

4.1.5. *Baseflow method*

The applied rainfall-runoff submodel includes an exponential recession model to represent the time variation of baseflow [26] (**Figure 3**).

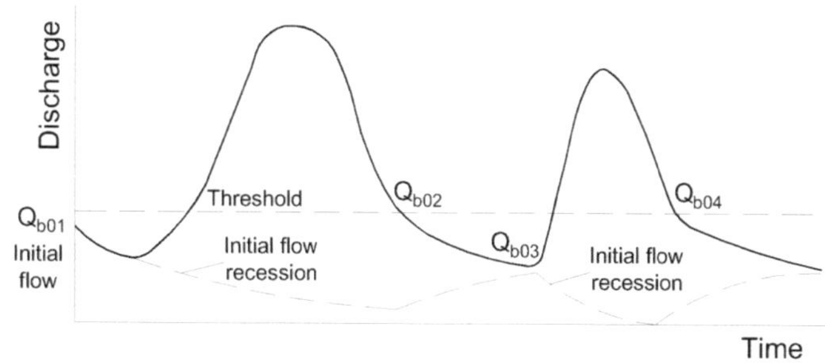

Figure 3. Recession with multiple runoff peaks.

The baseflow recession model can be described by Eq. (10).

$$Q_{bt} = Q_{b0}\,k^{t} \tag{10}$$

where Q_{bt} is the baseflow at any time t, Q_{b0} is the initial baseflow (at time zero) and k is the exponential decay constant; it is defined as the ratio of the baseflow at time t to the baseflow one day earlier.

4.1.6. Routing model

For the routing of the total discharge (direct runoff + baseflow) hydrograph in the main streams of the basin, the Muskingum-Cunge model is used. The Muskingum-Cunge model is based on the widely used hydrologic routing Muskingum model [27].

The aforementioned methods are incorporated and are given as options in the "sub-basin" and "reach" editors in the hydrologic model HEC-HMS 4.2 [28]. The result from HEC-HMS is the development of runoff, baseflow and total discharge hydrographs.

A more thorough outline of the rainfall-runoff submodel can be found in Ref. [29].

4.2. Soil erosion submodel

The objective of this submodel is to estimate the sediment yield that results from the erosion of the soil surface and reaches the main streams of the sub-basins. This is done by calculating firstly the amount of soil detachment by rainfall (detachment by raindrop impact) and secondly the amount of soil erosion due to surface runoff (shearing force of flowing water).

The estimation of soil erosion due to rainfall was performed by means of Poesen's relations [16]:

$$q_{rs} = C(KE)\,r_s^{-1}\cos a \tag{11}$$

$$q_r = q_{rs}\,[\,0.301\sin a + 0.019\,D_{50}^{-0.22}(1 - e^{2.42\sin a})\,] \tag{12}$$

where q_{rs} is the mass of detached particles per unit area (kg/m^2), C is the soil cover factor, KE is the rainfall kinetic energy (J/m^2), r_s is the soil resistance to drop detachment (J/kg), a is the slope gradient (°), q_r is the downslope splash transport per unit width (kg/m) and D_{50} is the median particle diameter (m).

The soil erosion due to runoff is calculated using Nielsen's relation [30]:

$$q_f = r_e q_t \tag{13}$$

where q_f is the sediment transport per unit width by runoff [m^3/(s m)], r_e is the entrainment ratio ($r_e = 1$ for non-cohesive soils, $r_e < 1$ for cohesive soils) and q_t is the sediment transport capacity per unit width by runoff [m^3/(s m)].

For the surface runoff sediment transport capacity, q_t, the modified formula of Engelund-Hansen is used [31]:

$$q_t = 0.04 \frac{(2g/f)^{1/6}}{[(\rho_s/\rho - 1)^2 g^{1/2} D_{50}]} q^{5/3} s^{5/3} \tag{14}$$

The sediment yield that reaches the stream is calculated by means of comparison between the available sediment and the sediment transport capacity by runoff [32].

4.3. Sediment transport submodel for streams

The final stage in the hydro-geomorphological modelling is the routing of sediments into the main streams. This is done on the basis of the comparison between the sediment transport capacity by streamflow and the available sediment in the stream. The available sediment in the stream is the sediment yield that was calculated by the soil erosion submodel. This comparison between sediment transport capacity by streamflow and the available sediment in the stream defines whether erosion or deposition takes place in the stream.

Sediment transport capacity by streamflow is estimated from the sediment concentration in the stream, which is computed by the unit stream power model of Yang and Stall [17]:

$$\log c_{ts} = 5.435 - 0.286 \log \frac{w D_{50}}{v} - 0.457 \log \frac{u_*}{w} + \left(1.799 - 0.409 \log \frac{w D_{50}}{v} - 0.314 \log \frac{u_*}{w}\right) \log\left(\frac{us}{w} - \frac{u_{cr} s}{w}\right) \tag{15}$$

$$\frac{u_{cr}}{w} = \frac{2.5}{\log(u_* D_{50}/v - 0.06)} + 0.66, \quad \text{if } 1.2 < \frac{u_* D_{50}}{v} < 70 \tag{16}$$

$$\frac{u_{cr}}{w} = 2.05, \quad \text{if } \frac{u_* D_{50}}{v} \geq 70 \tag{17}$$

where c_{ts} is the total sediment concentration by weight (ppm), w is the terminal fall velocity of sediment particles (m/s), D_{50} is the median particle diameter (m), v is the kinematic viscosity of the water (m^2/s), s is the energy slope, u is the mean flow velocity (m/s), u_{cr} is the critical mean flow velocity (m/s) and u_* is the shear velocity (m/s).

5. Application and calibration of the model

Once all the necessary data have been gathered and processed, the model is applied. This is the first essential step of modelling. This procedure should be followed by the calibration of the model and finally by its validation. The integrated procedure is schematically described in **Figure 4**.

http://dx.doi.org/10.5772/intechopen.68655

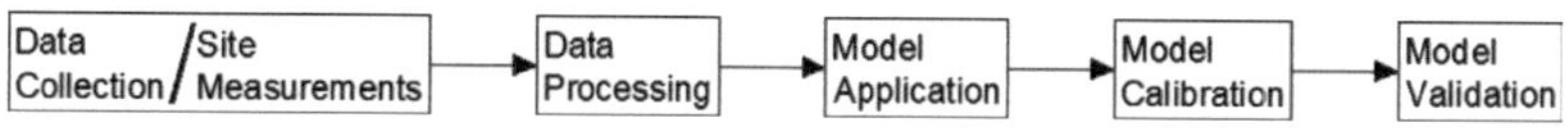

Figure 4. Schematic representation of modelling process.

5.1. Kosynthos river basin

The hydrologic model was applied for the data period of 2005–2009, that is from 1 January 2005 to 15 March 2009, in HEC-HMS 4.2. The resulting runoff, baseflow and total discharge hydrographs, from HEC-HMS, along with the rainfall amount, were used as the decisive input data to the soil erosion model, the outcome of which is the sediment inflow to main streams. Subsequently, the sediment available in the main streams, in combination with the sediment transport capacity of the main streams, determines the final sediment yield at the outlet of the basin.

The continuous nature of the hydrograph enables the development of a continuous sediment graph. For more precise calculations, the basin was divided into 10 natural sub-basins (**Figure 1**).

All the submodels of the composite mathematical model were calibrated. A simple calibration procedure was followed. All submodels ran with the calibrated parameters starting with their initial values. For every new run of the CMM, the calibrated parameters reentered the submodels with a changed value. This procedure was applied repeatedly until computations and measurements reached the best possible agreement. The calibrated parameters are shown in **Table 1**. The numbers in brackets refer to either the percentage change or the final value obtained by the parameters, whilst the numbers in parentheses refer to the times that each parameter has been reentered in the model, with a changed value, during the calibration process.

Recession constant and threshold discharge are two parameters used in HEC-HMS to reset baseflow during a rainfall event. These two parameters are very crucial, as they greatly regulate baseflow and, therefore, total stream discharge.

The recession constant describes the rate at which baseflow recedes between storm events. It is defined as the ratio of baseflow at the current time to the baseflow 1 day earlier.

The entrainment ratio defines the cohesion of the soil and has been described in Section 4.2 [Eq. (13)].

Roughness coefficient is a parameter that greatly affects the transport capacity by streamflow and, hence, the final sediment discharge at the basin outlet.

Parameters that cannot be directly measured, in comparison with parameters that are measurable (rainfall depth, water discharge, meteorological data, etc.), are best suited for calibration. Such parameters, amongst others, can be the calibrated parameters in **Tables 1** and **2**, namely: curve number, lag time, baseflow recession constant, flow ratio, entrainment ratio and streambed roughness. Nevertheless, all parameters were considered for calibration within a reasonable range to ensure these do not lose their physical meaning.

5.2. Nestos river basin

For the needs of modelling, the basin was divided into 20 natural sub-basins (**Figure 2**). The model was applied separately for four different cases. For each case, each one of the four measurement sites (Paschalia, Stavroupoli, Galani, basin outlet) (**Figure 2**) was considered as the outlet of the basin. This implies the exemption of several sub-basins, for example, the case of Paschalia and Stavroupoli.

The difference between the simulations for Galani and the basin outlet is that Galani is located upstream of the irrigation canals, whilst the outlet of the basin is located slightly downstream. Hence, in the case of the basin outlet, the discharge allocated for irrigation was extracted from the total discharge, during the irrigation periods.

The discharges of Platanovrysi dam were inputted at a daily time step in HEC-HMS. As far as the operation of HEC-HMS is concerned, the dam is regarded as a source point.

Submodels	Calibrated parameters			
Rainfall-runoff submodel	CN	Lag time	Baseflow recession constant	Flow ratio
	[+5%]	[−3%]	[0.9]	[0.27]
	(7)	(6)	(4)	(4)
Soil erosion submodel (relationships of Poesen)	Entrainment ratio			
	[0.65]			
	(6)			
Stream sediment transport submodel	Streambed roughness			
	[−10%]			
	(8)			

Table 1. Calibrated parameters for Kosynthos river basin.

Submodels	Calibrated parameters			
Rainfall-runoff submodel	CN	Lag time	Baseflow recession constant	Flow ratio
	[−8%]	[+10%]	[0.81]	[0.16]
	(9)	(8)	(6)	(6)
Soil erosion submodel (relationships of Poesen)	Entrainment ratio			
	[0.7]			
	(8)			
Stream sediment transport submodel	Streambed roughness			
	[+15%]			
	(6)			

Table 2. Calibrated parameters for Nestos river basin.

All the simulations for Nestos river basin cover a period from 11 September 2005 to 31 July 2014.

A multi-site calibration procedure was implemented for Nestos river basin. The two upstream measurement sites, of Paschalia and Stavroupoli, were used for calibration of the model, whilst the measurements of the two downstream measurement sites, of Galani and the basin outlet, were used for the validation of it. The calibrated parameters are the same with those for Kosynthos river basin and are shown in **Table 2**. At this point, it has to be noted that the calibration of the CMM was dealt with two different objectives for the two basins. This explains the different values obtained by the calibrated parameters in **Tables 1** and **2**. For instance, the increase of discharge was sought in the case of Kosynthos river basin, whilst for Nestos river basin the decrease of discharge was intended.

6. Results: model validation

6.1. Kosynthos river basin

6.1.1. Water discharge

In **Figure 5**, the computed discharge hydrograph (direct runoff + baseflow) and the measured discharge values at the basin outlet are depicted for the period October 2005–March 2009.

Discharge is mainly driven by rainfall. This means that in high rainfall seasons there is a discharge increase. From the above hydrograph, the months with the highest rainfall depths and discharges are March, October, November and December.

6.1.2. Sediment discharge

In **Figures 6–8**, the computed sediment graphs (bed load + suspended load) at the basin outlet are illustrated for the periods October 2005–July 2006, April–May 2007 and April 2008–January 2009.

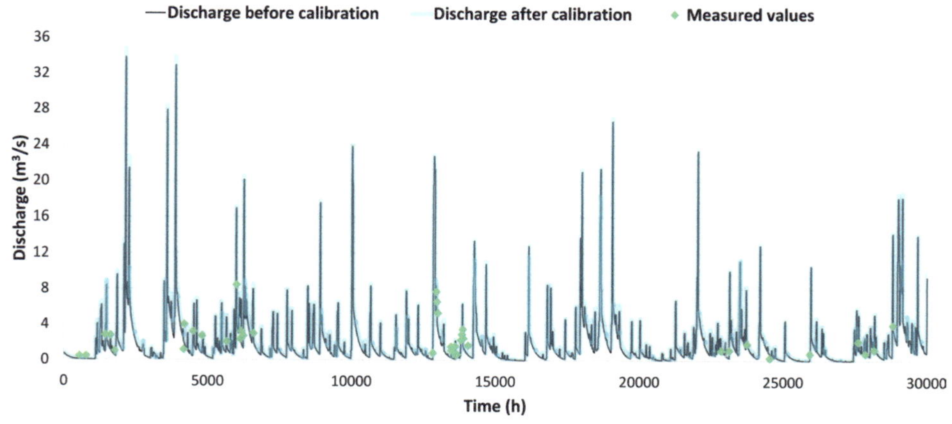

Figure 5. Discharge hydrograph (October 2005–March 2009, Kosynthos river basin outlet).

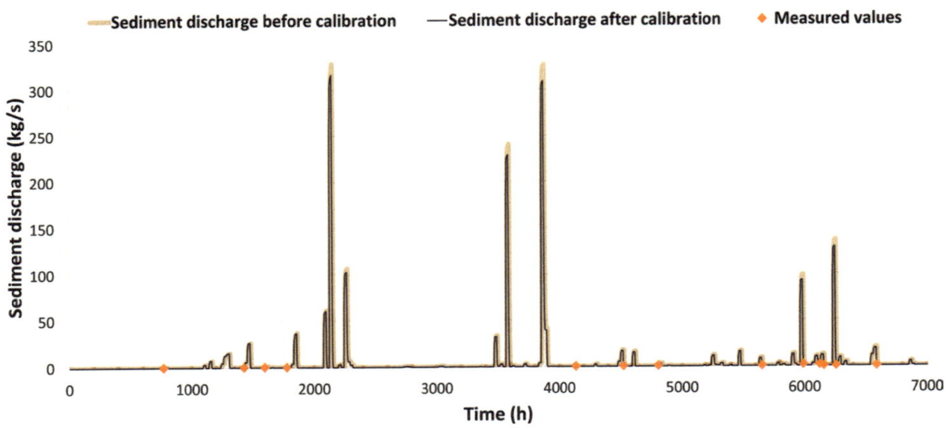

Figure 6. Sediment graph (October 2005–July 2006, Kosynthos river basin outlet).

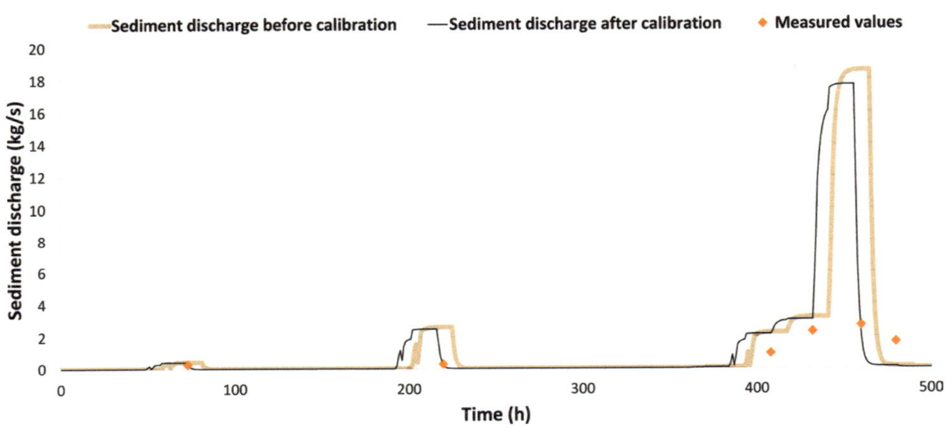

Figure 7. Sediment graph (April–May 2007, Kosynthos river basin outlet).

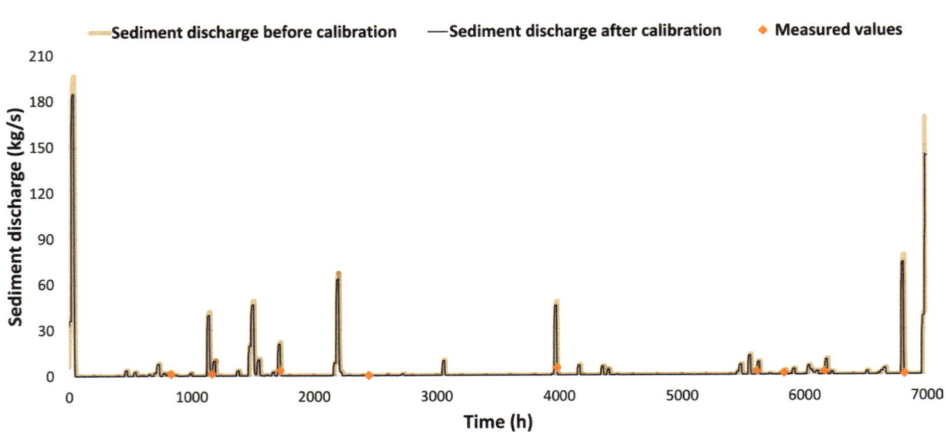

Figure 8. Sediment graph (April 2008–January 2009, Kosynthos river basin outlet).

As already mentioned, sediment discharge is mainly due to soil erosion, which in turn is a function of rainfall and runoff. Hence, the sediment discharge should be expected to be high when the discharge is high. Indeed, the sediment graphs follow a similar pattern with that of the hydrographs. High sediment discharges are observed in the same months as high water discharges, that is March, October, November and December. Sediment discharge also presents high values in January and May.

6.2. Nestos river basin

6.2.1. Water discharge

In **Figures 9** and **10**, the comparison between the computed discharge hydrographs for the two measurement sites used for the calibration of the model, Paschalia and Stavroupoli, before and after the calibration, is depicted. In **Figure 11**, the computed discharge hydrograph and the measured discharge values at Galani measurement site are depicted for the period March

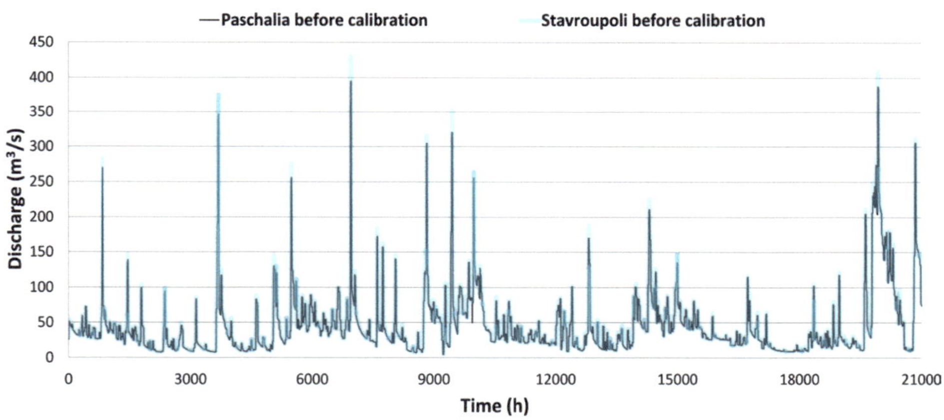

Figure 9. Discharge hydrograph (October 2006–March 2009, Paschalia-Stavroupoli before calibration).

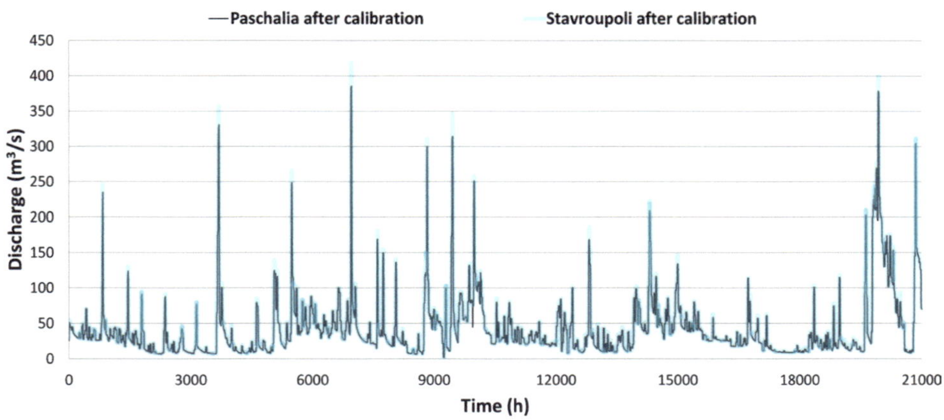

Figure 10. Discharge hydrograph (October 2006–March 2009, Paschalia-Stavroupoli after calibration).

2006–January 2009. In **Figures 12–15**, the computed discharge hydrographs and the measured discharge values at the basin outlet are depicted for the periods September–October 2005, July–November 2008, July 2009–January 2010 and March–May 2011.

As it can be observed from **Figures 9** and **10**, the hydrographs of the two upstream measurement sites follow the very same pattern, with exception that the discharge at Paschalia, which is located a little upstream from Stavroupoli, is slightly lower. Further downstream, at Galani measurement site, the discharges increase even more. This is due to the addition of more sub-basins and their hydrological distribution to Nestos River. Discharges higher than 200 m^3/s are mostly observed in months January, March, October, November and December at the aforementioned measurement sites.

The case is different at the basin outlet, which is the final measurement site. Here, the presence of the two irrigation canals, which divert large volumes of water during the irrigation period, is a game-changing element as far as the discharge is concerned. The two canals irrigate the plains of Kavala and Xanthi. The irrigation period, in both cases, starts in April and goes up to

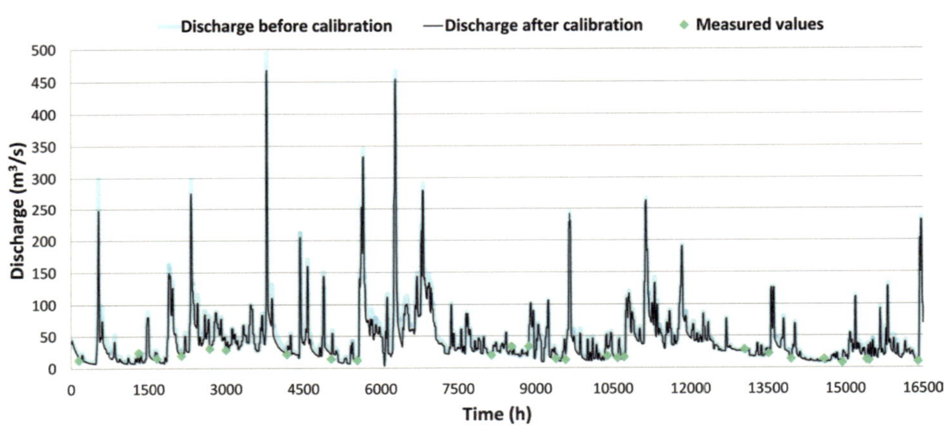

Figure 11. Discharge hydrograph (March 2006–January 2009, Galani).

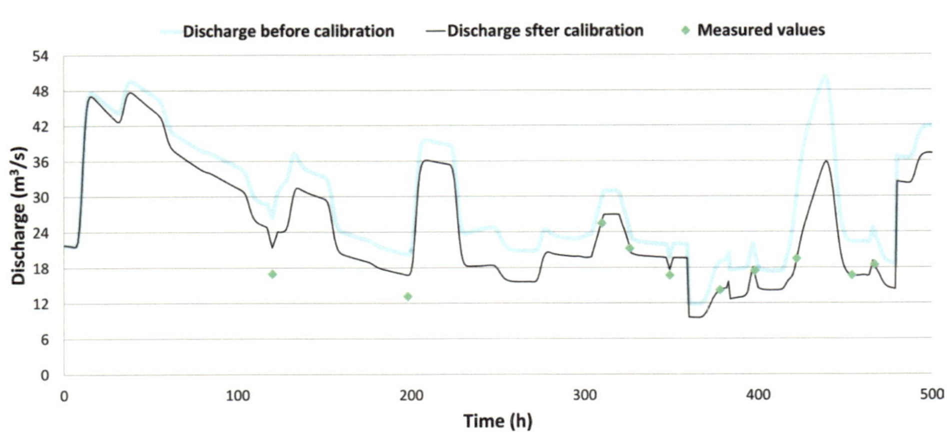

Figure 12. Discharge hydrograph (September–October 2005, Nestos river basin outlet).

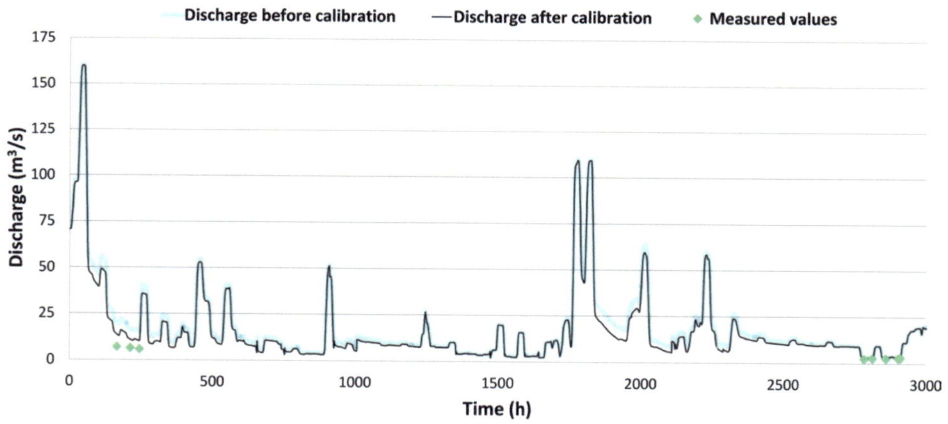

Figure 13. Discharge hydrograph (July–November 2008, Nestos river basin outlet).

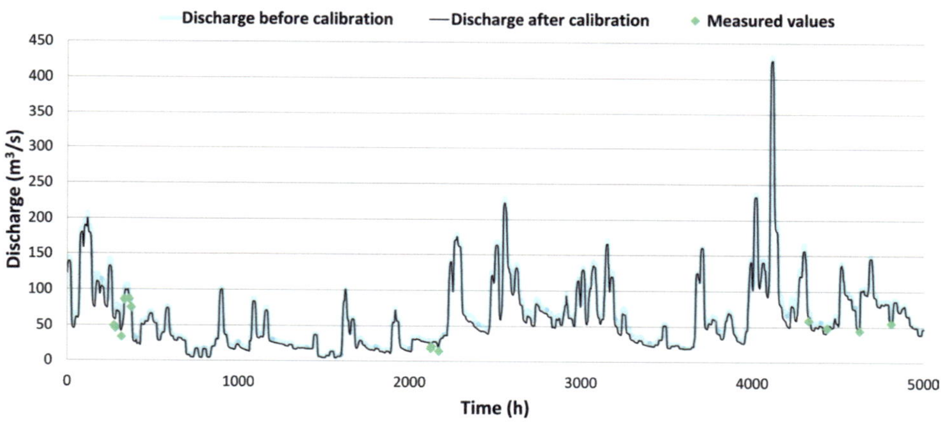

Figure 14. Discharge hydrograph (July 2009–January 2010, Nestos river basin outlet).

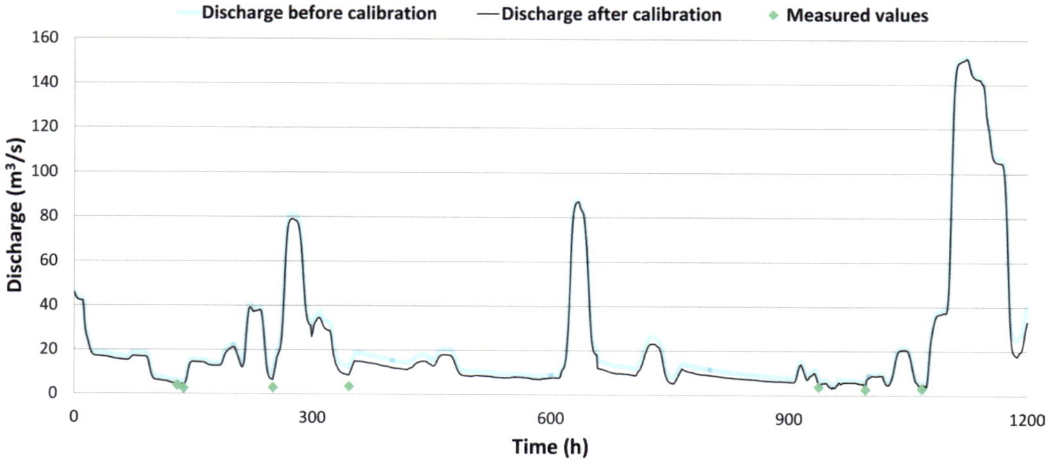

Figure 15. Discharge hydrograph (March–May 2011, Nestos river basin outlet).

October. The amount of water that gets subtracted from Nestos River is much higher during the drier months of the summer. Especially during this period, the Platanovrysi dam holds a vital role in the sustainability of the water resources management by replenishing the riverine system and by securing environmental flows.

6.2.2. Sediment discharge

In **Figures 16–18**, the computed sediment graphs (bed load + suspended load) at the basin outlet are illustrated for the periods September–October 2005, June 2008–June 2011 and October 2011–July 2014.

Despite the fact that the Platanovrysi dam hinders a portion of sediment, the basin itself provides the streams with the necessary amount of sediment that is vital for the aquatic ecosystems and for a variety of environmental reasons. The simulated sediment discharge reaches an overall average value of approximately 30 kg/s, whilst—as seen from **Figures 17** and **18**—there are several high peaks as well.

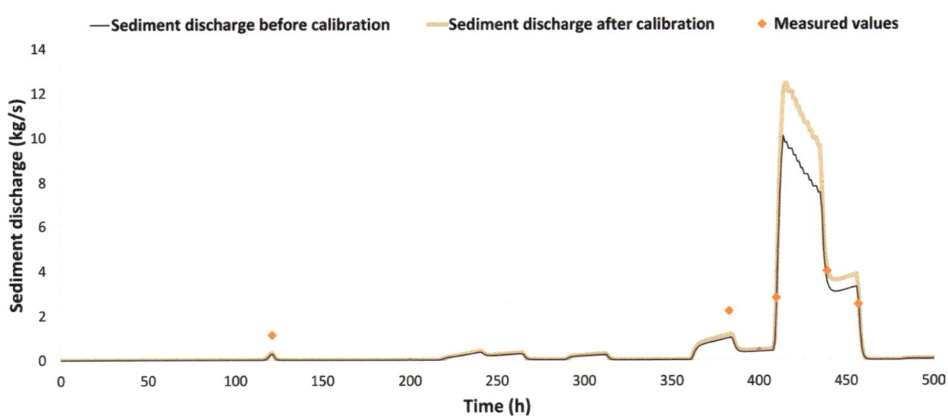

Figure 16. Sediment graph (September–October 2005, Nestos river basin outlet).

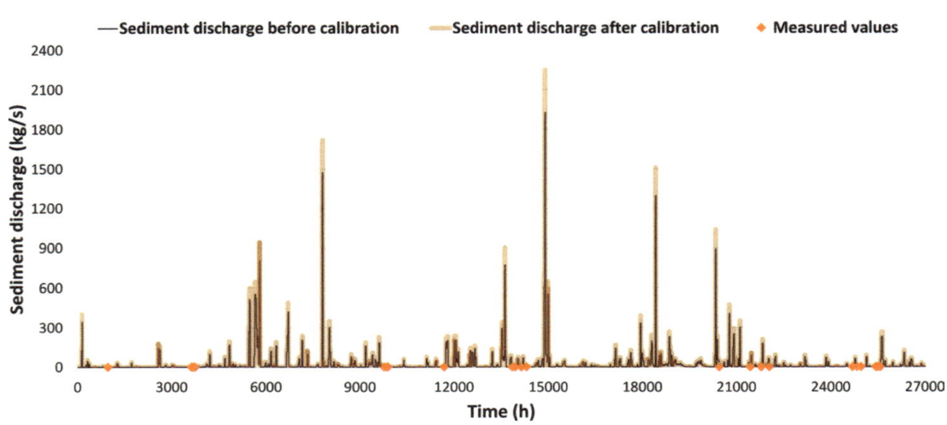

Figure 17. Sediment graph (June 2008–June 2011, Nestos river basin outlet).

http://dx.doi.org/10.5772/intechopen.68655

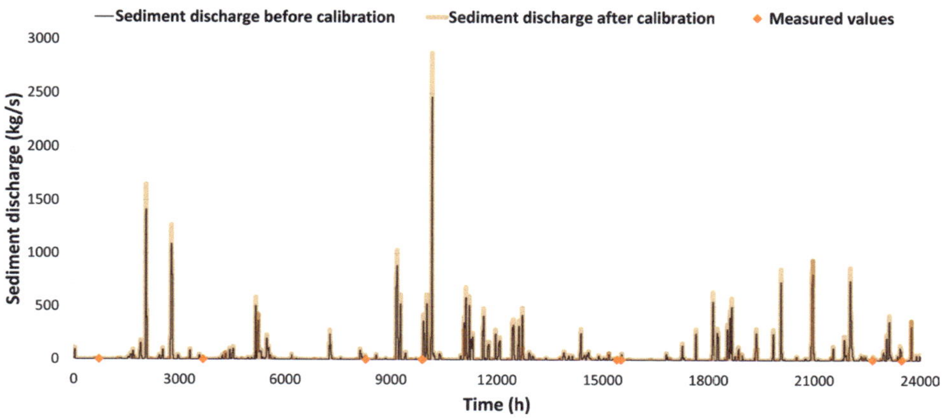

Figure 18. Sediment graph (October 2011–July 2014, Nestos river basin outlet).

6.3. Efficiency criteria

A series of efficiency criteria were used for the comparison between computed and measured water discharge and sediment discharge values. The criteria utilized, as well as their values, are shown in **Table 3**.

Efficiency criteria	Kosynthos river basin		Nestos river basin		
	Basin outlet		Galani	Basin outlet	
	Water discharge	Sediment discharge	Water discharge	Water discharge	Sediment discharge
Mean absolute error (m^3/s, kg/s)	0.4062	0.4239	2.9332	2.4944	0.2876
Mean relative error (%)	7.5528	6.9342	20.0945	25.4658	12.3819
Mean square error (m^6/s^2, kg^2/s^2)	0.4984	0.5358	19.5876	14.9494	0.3829
Root mean square error (m^3/s, kg/s)	0.706	0.732	4.4258	3.8665	0.6188
PBIAS (%)	−1.5411	−11.7108	−15.4909	−10.3444	12.3711
Nash-Sutcliffe efficiency	0.8641	0.5899	0.65	0.9737	0.6469
Index of agreement, *d*	0.9649	0.9096	0.9103	0.9936	0.8917
Coefficient of persistence	0.9006	0.7413	0.6627	0.9603	0.6413
Coefficient of performance	0.1359	0.4101	0.35	0.0263	0.3531
Correlation coefficient	0.9318	0.8485	0.8943	0.9925	0.8127
Determination coefficient	0.8682	0.7199	0.7997	0.985	0.6605
t-Test (*P*-value > 0.05)	0.9355	0.6781	0.1926	0.675	0.6866

Table 3. Efficiency criteria values.

The validation of a time series model, such as the one applied in the current study, is integrated by the use of appropriate statistic criteria of efficiency, which are applied for the comparison between calculated and measured values and determine whether a model is successful or not.

There is a wide range of literature on the efficiency criteria utilized in this study. Their formulas, theoretical concepts and efficiency ranges (**Table 4**) should be comprehended before one uses them.

Efficiency criteria	Range of efficiency
Mean absolute error (m^3/s, kg/s)	The closer to 0 (perfect fit) the MAE is, the closer the fit between observed and predicted values
Mean relative error (%)	The closer to 0 (perfect fit) the MRE is, the closer the fit between observed and predicted values. A (−) or a (+) indicates an underprediction or an overprediction of the model
Mean square error (m^6/s^2, kg^2/s^2)	The closer to 0 (perfect fit) the MSE is, the closer the fit between observed and predicted values
Root mean square error (m^3/s, kg/s)	The closer to 0 (perfect fit) the RMSE is, the closer the fit between observed and predicted values
PBIAS (%)	(−) indicates an overprediction of the model, whilst (+) indicates an underprediction of the model. The optimal value is (0). Satisfactory values according to Moriasi et al. [33]: streamflow = ±25% and sediment = ±55%
Nash-Sutcliffe efficiency	Efficiencies range from −Inf to 1, with 1 being the optimal value (perfect fit)
Index of agreement, *d*	The index of agreement ranges from 0 to 1, with 1 being the optimal value (perfect fit)
Coefficient of persistence	C_P ranges from 0 to 1, with 1 being the optimal value, whilst a value larger than 0 indicates a "minimally acceptable" model performance [34]
Coefficient of performance	The coefficient of performance approaches to 0 as the predicted values approach the observed ones
Correlation coefficient, *r*	$-1 \leq r \leq +1$. The + and − signs are used for positive and negative linear correlations, respectively. A perfect correlation of ± 1 occurs when all the data points lie exactly on a straight line
Determination coefficient, r^2	The range of r^2 lies between 0 and 1. A value of zero means no correlation at all, whereas a value of 1 means that the dispersion of the prediction is equal to that of the observation
t-Test (*P*-value > 0.05)	The acceptance or not of a *t*-test's result relies on whether a probability value (*P*-value) is greater or not than the significance level (alpha). Alpha was set to 0.05

Table 4. Efficiency ranges of the statistic criteria.

7. Discussion

The hydrologic model was calibrated under four parameters which regulate the surface runoff, as well as the baseflow and the total discharge. The decrease or the increase of

CN "constitutes" the soil surface more or less susceptible to infiltration by water and, therefore, leads to a proportionate reduction or increment of surface runoff. An increase of lag time enhances the infiltration process by "retaining" runoff for more time on the soil surface, augmenting the hydrologic losses this way. The decrease of lag time has the opposite effect.

The most influencing parameters of HEC-HMS, as far as baseflow and total stream discharge are concerned, were proved to be the baseflow recession constant and the flow ratio. These two parameters have a high and direct impact on baseflow and total discharge. However, they also constitute an indirect influence on the sediment transport in streams by affecting the stream sediment transport capacity.

Both sediment discharge and sediment yield are highly dependent on surface runoff. Surface runoff is the decisive parameter in soil erosion; the lower the runoff discharge is, the lower is the erosive force and the sediment transport capacity of the flowing water.

Soil cohesion plays a decisive role in soil erosion, hence, the calibration of the entrainment ratio. Generally, the lower the cohesion of the soil, the more easily the soil particles are rived and eroded.

A decrease of the streambed roughness coefficient leads to a decrease of the stream sediment transport capacity. This leads to channel deposition which, in turn, results in a decrease of sediment discharge at the basin outlet.

In the past, the annual sediment yield because of rainfall and runoff was computed at the outlet of Kosynthos river basin [35] and at the outlet of Nestos river basin [32]. However, in both cases, the computations were performed on a monthly basis.

8. Conclusions

It is concluded that there is a very good approximation between the computed and measured discharge and sediment discharge values and that the deviation between these values is not considerable.

According to Ref. [36], the full benefit of an erosion prediction model is gained through the use of a continuous simulation model. By continuous simulation, it is meant that the model follows the time variation of the physical processes related to erosion.

The combination of a hydrologic model with a soil erosion model and a stream sediment transport model enables the transition of the hydrograph, due to a rainfall event, at a basin outlet to the corresponding sediment graph. In other words, the variation with time of the sediment discharge at the basin outlet is computed on the basis of the variation with time of the stream discharge due to the rainfall event [37]. The continuous nature of the hydrograph enables the development of a continuous sediment graph. It could be stated that the connection of the models through the input-output data ensures the almost parallel running of the models, which approximates the natural reality.

It is believed that the deviations between computed and measured water discharge and sediment discharge values for single rainfall events can be mitigated by means of the continuous hydro-geomorphologic modelling because of the integrating effect obtained through the use of a long simulation period. Continuous hydro-geomorphologic modelling for a relatively long time period, in a relatively large basin, provides a more realistic representation of the runoff process, as well as the soil erosion and sediment transport processes.

Seeing that the overwhelming majority of rivers in Greece are ungauged, HEC-HMS can reliably be used to calculate stream discharges and simulate river flows across the country. It can also be applied to other parts of the Mediterranean, with similar morphological and climatic conditions. Additionally, given the fact that the process of measuring the sediment discharge is a difficult and laborious task—yet of great significance for a variety of reasons—the composite mathematical models presented here can be used in the Greek mountainous terrain to successfully estimate soil erosion, sediment discharges and sediment yields at the basin scale.

Finally, continuous simulation of the hydromorphological processes can be an indicator of the future trends concerning their quantification.

The results of the efficiency criteria conclude that the continuous hydro-geomorphologic modelling can be successfully applied to both Kosynthos river basin and Nestos river basin.

Author details

Konstantinos Kaffas* and Vlassios Hrissanthou

*Address all correspondence to: kostaskaffas@gmail.com

Department of Civil Engineering, Democritus University of Thrace, Kimmeria Campus, Xanthi, Greece

References

[1] Beasley DB, Huggins LF, Monke EJ. ANSWERS—A model for watershed planning. Transactions of the ASAE. 1980;**23**:938-944

[2] Merritt WS, Letcher RA, Jakeman AJ. A review of erosion and sediment transport models. Environmental Modelling & Software. 2003;**18**(8):761-799

[3] Holtan HN. A Concept for Infiltration Estimates in Watershed Engineering. Washington, DC: USDA-ARS Bulletin. 1961; pp. 41-51.

[4] Green WH, Ampt CA. Studies on soil physics, I. Flow of water and air through soils. Journal of Agricultural Science. 1911;**4**:1-24

[5] Wischmeier WH, Smith DD. Predicting Rainfall Erosion Losses: A Guide to Conservation Planning. In: Agriculture Handbook 282. 1400 Independence Ave., S.W., Washington, DC: USDA-ARS; 1978

[6] Yalin MS. An expression for bed load transportation. Journal of Hydraulics Division, ASCE. 1963;**98**(HY3):221-250

[7] Foster GR, Meyer LD. Transport of particles by shallow flow. Transactions of the ASAE. 1972;**15:**99-102

[8] Alonso CV, Nielberg WH, Foster GR. Estimating sediment transport capacity in watershed modeling. Transactions of the ASAE. 1981;**24:**1211-1220

[9] Young RA, Onstad CA, Bosch DD, Anderson WP. AGNPS, agricultural nonpoint source pollution. A watershed analysis tool. In: Conservation Research Report 35. Washington, DC: USDA; 1987

[10] Young RA, Onstad CA, Bosch DD, Anderson WP. AGNPS: A nonpoint-source pollution model for evaluating agricultural watersheds. Journal of Soil and Water Conservation. 1989;**44**(2):4522-4561

[11] Soil Conservation Service (SCS). National Engineering Handbook, Section 4: Hydrology. Washington, DC: USDA; 1972

[12] Renard KG, Foster GR, Weesies GA, McCool DK, Yoder DC. Predicting Soil Erosion by Water: A Guide to Conservation Planning with the Revised Universal Soil Loss Equation (RUSLE), Vol. 703. Washington, DC: US Government Printing Office; 1977

[13] Einstein HA. The Bed-Load Function for Sediment Transportation in Open Channel Flows. In: Technical Bulletin 1026. USDA; 1950

[14] Bagnold RA. Bedload transport in natural rivers. Water Resources Research. 1977;**13**(2): 303-312

[15] Knisel WG. CREAMS: A Field-scale Model for Chemicals, Runoff, and Erosion from Agricultural Management System. In: Conservation Research Report No. 26. Washington, DC: USDA-SEA; 1980

[16] Poesen J. An improved splash transport model. Zeitschrift für Geomorphologie. 1985;**29**: 193-211

[17] Yang CT, Stall JB. Applicability of unit stream power equation. Journal of the Hydraulics Division, ASCE. 1976;**102**(5):559-568

[18] Paschalidis G, Iordanidis I, Anagnostopoulos P. Discharge and sediment transport in a basin with a dam at its upper boundary. In: Proceedings of the 12th International Conference on Protection and Restoration of the Environment; 29 June–3 July 2014; Skiathos Island, Greece

[19] Global Weather Data for SWAT [Internet]. Available from: http://globalweather.tamu.edu/. [Accessed: 13 December, 2016]

[20] Thiessen AH. Precipitation averages for large areas. Monthly Weather Review. 1911;**39**(7): 1082-1089

[21] Allen RG, Pereira LS, Raes D, Smith M. Crop evapotranspiration: Guidelines for computing crop requirement, irrigation and drainage, Paper No. 56. Rome: FAO; 1998

[22] Giandotti M. Forecast from full and lean courses of water. Ministry LL.PP., Hydrographic Survey, Vol. 8, Rep. No. 2; 1934. Hydrographic Department of Italy, Rome (in Italian)

[23] Giandotti M. Empirical flood forecasting based on the meteoric precipitations, the physical and morphological characteristics of the basins; application of the method to some of the Ligurian basins. Memorie e Studi Idrografici. 1940;**10**:5-13 (in Italian)

[24] Pasini F. Report on the renana recovery plan. 1914; Bologna, Italy (in Italian)

[25] Soil Conservation Service (SCS). National Engineering Handbook, Section 4: Hydrology. Springfield VA: USDA; 1971

[26] Barnes BS. The structure of discharge recession curves. Transactions of the American Geophysical Union. 1939;**20**(4):721-725

[27] Cunge JA. On the subject of a flood propagation computatronal method (Muskingum method). Journal of Hydraulic Research. 1969;**7**(2):205-230

[28] United States Army Corps of Engineers (USACE). Hydrologic Modeling System HEC-HMS, User's Manual, Version 4.0. 2013; Hydrologic Engineering Center, Davis, CA

[29] Kaffas K, Hrissanthou V. Application of a continuous rainfall-runoff model to the basin of Kosynthos river using the hydrologic software HEC-HMS. Global NEST Journal. 2014;**16**(1):188-203

[30] Nielsen SA, Storm B, Styczen M. Development of distributed soil erosion component for the SHE hydrological modelling system. In: Proceedings of the International Conference on Water Quality Modelling in the Inland Natural Environment, BHRA; June 1986; Bournemouth, UK. pp. 1-13

[31] Engelund F, Hansen E. A monograph on sediment transport in alluvial streams. Copengahen: Teknisk Forlag; 1967

[32] Hrissanthou V. Comparative application of two erosion models to a basin. Hydrological Sciences-Journal-des Sciences Hydrologiques. 2002;**47**(2):279-292

[33] Moriasi DN, Arnold JG, Van Liew MW, Bingner RL, Harmel RD, Veith TL. Model evaluation guidelines for systematic quantification of accuracy in watershed simulations. Transactions of the ASABE. 2007;**50**(3):885-900

[34] Gupta HV, Sorooshian S, Yapo PO. Status of automatic calibration for hydrologic models: Comparison with multilevel expert calibration. Journal of Hydrologic Engineering. 1999;**4**(2):135-143

[35] Hrissanthou V, Delimani P, Xeidakis G. Estimate of sediment inflow into Vistonis Lake, Greece. International Journal of Sediment Research. 2010;**25**(2):161-174

[36] Lal R. Soil Erosion Research Methods. Soil and Water Conservation Society and St. Lucie Press, 6000 Broken Sound Parkway, NW, Boca Raton, FL; 1994

[37] Hrissanthou V, Theodorakopoulos E. Computation of sediment graph for a flood event. In: Proceedings 6th International Conference of the European Water Resources Association (EWRA); 2005; Menton, France; pp. 1-13

Chapter 3

Analysis of the Impacts of Changes in Streamflow and of Restoration on the Morphological Evolution of the Matambin River Channel in the St. Lawrence Lowlands (Quebec, Canada)

Ali Assani, Lisanne Chauvette and
Stéphane Campeau

Additional information is available at the end of the chapter

http://dx.doi.org/10.5772/intechopen.68444

Abstract

Although many plots of land that were once farmed have been reforested in the St Lawrence Lowlands of Quebec since the 1950s, no study has yet looked at the morphological impacts of this land-use change. To address this, we analyzed the evolution of the Matambin River (99 km^2) channel width and sinuosity using diachronic analysis of air photographs taken between 1935 and 2008. Results of this analysis show a roughly 21% decrease in mean channel bankfull width from 1935 to 1964. This time interval was characterized by a low frequency of strong flood flows in the region and a roughly 32% increase in the forested land area, the reforestation having started in the 1950s. After 1964, a trend of increasing mean channel bankfull width is observed. This increase is associated with the increase in frequency of strong flood flows in the region and a decrease in the amount of suspended sediments produced by soil erosion following the increase in forest cover in the watershed. In contrast, channel sinuosity did not change much over the period from 1935 to 2008.

Keywords: reforestation, streamflow, air photographs, width, sinuosity, Matambin, Quebec

1. Introduction

Unlike fluvial processes, the temporal evolution channel morphology in Quebec remains poorly studied. A few recent studies have looked at the impacts of dams on this evolution [1–3], and have highlighted the influence of flow management mode and lithology on the morphological evolution of channels downstream from dams and reservoirs. In the

INTECH
open science | open minds

© 2017 The Author(s). Licensee InTech. This chapter is distributed under the terms of the Creative Commons Attribution License (http://creativecommons.org/licenses/by/3.0), which permits unrestricted use, distribution, and reproduction in any medium, provided the original work is properly cited. (cc) BY

Canadian Shield, where rivers are characterized by nearly regular sequences of alternating sandy and boulder reaches, the morphological impacts of an inversion of the natural cycle of flows (maximum flows occurring in winter and minimum flows in springtime during snowmelt) result in a trend of decreasing channel bankfull width in boulder reaches. Changes observed in sandy reaches are characterized by the multiplication and growth of islets and banks [3]. In the St Lawrence Lowlands, which are mainly underlain by unconsolidated deposits, dams lead to an increase in mean bankfull width, but a decrease in channel sinuosity [2].

As for the impacts of agriculture, which is restricted to the St Lawrence Lowlands, on the morphological evolution of channels, it is well established that drainage work done in the 1950s and 1960s led to a linearization of channels due to the artificial cutoff of numerous meanders, this linearization leading to channel widening. However, after many decades of intensive farming that resulted in massive watershed deforestation, gradual reforestation of previously farmed plots has been observed since the 1950s. This phenomenon is also observed in many other developed countries as a result of declining farming populations (e.g., [4–6]). Few studies have yet looked at the impacts of this land-use change (reforestation) on the morphological evolution of channels (e.g., [4, 6–16]) compared to studies looking at its hydrological impacts [17]. These different studies have shown that in addition to vegetation-related factors (age and type of trees, surface area covered, presence or absence of herbaceous strata, etc.), the impacts of reforestation also depend on the evolution of rivers flows and sediment grain size.

The impacts of reforestation on the morphological evolution of channels have not yet been studied in Quebec. Because of the growing increase in reforested surface area in many watersheds in recent decades, it is important to constrain the impacts of this phenomenon on the morphological evolution of channels. The goal of this study is to constrain the impacts of reforestation and hydroclimate variability on the morphology (width and sinuosity) of the Matambin River channel in Quebec.

2. Methodology

2.1. Description of the Matambin River watershed

The total surface area of the Matambin River watershed is 99 km^2. The river originates in the Canadian Shield and flows out into Lake Maskinongé, which is an outgrowth of the Maskinongé River, a tributary of the St Lawrence River. The Matambin River flows in large part through the St Lawrence Lowlands, which are characterized by a very low slope. It is relatively small watershed, its sinuous course, and the presence of farmlands (no major urban area or dam) are the main reasons for selecting this river. In addition, air photographs dating back to 1935 are available, allowing a diachronic analysis of changes in forest cover and channel morphology (**Figures 1** and **2**).

Elevation in the Matambin River watershed ranges from 135 to 500 m, the elevation of more than 80% of the watershed being less than 200 m within the St Lawrence Lowlands. The

http://dx.doi.org/10.5772/intechopen.68444

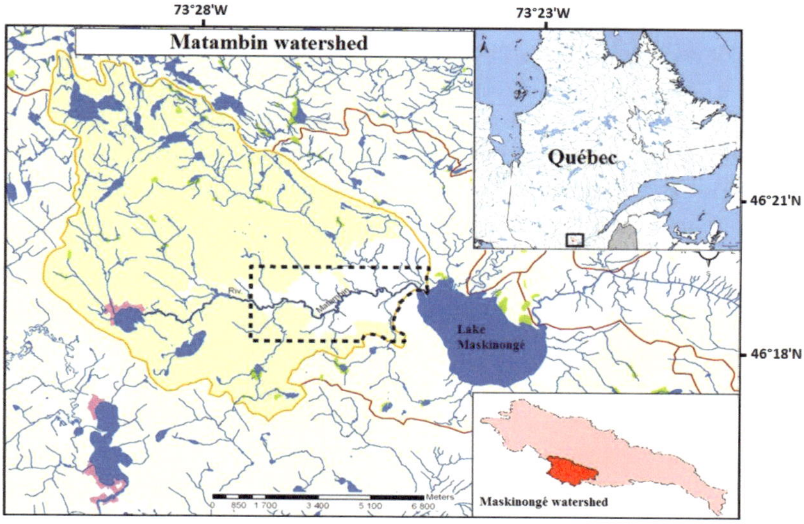

Figure 1. Location of the Matambin River watershed.

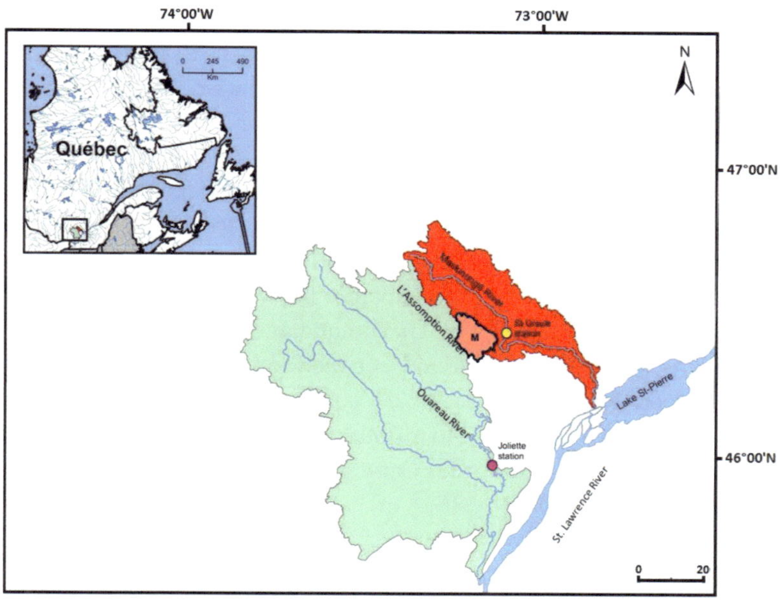

Figure 2. Location of the two water-gauging stations. M = Matambin River watershed.

Matambin River flows through undifferentiated glacial and fluvio-glacial alluviums, then downstream through lacustrine-marine deposits, and finally through marine and deltaic alluviums and undifferentiated alluviums. Thus, the river flows through deposits mainly consisting of fine clayey, silty, and sandy sediments.

From a hydrographic standpoint, the Matambin River watershed comprises numerous lakes, including five major ones, with surface areas ranging from 0.23 (Lac Migué) to 1.21 km^2 (Lac Blanc). Other lakes include Lac Quesnel (0.37 km^2), Lac Corbeau (0.48 km^2), and finally, Lac Matambin (0.60 km^2), located at the head of the Matambin River. All these lakes are within the Canadian Shield, upstream from the agricultural zone studied. The river is also fed by very small streams within the Canadian Shield. The study area is characterized by a humid continental type climate with warm summers (particularly in July) and very cold winters. Due to topographic variations within the watershed, precipitations range from 800 to 1359 mm in the Maskinongé River watershed.

Although it is located within the St Lawrence Lowlands, and as a result of reforestation, roughly 80% of the Matambin River watershed surface area is covered by forests, primarily mixed and deciduous in nature. There is very little logging in the area. According to an overview of the Matambin River watershed [18], no hectare of forest was logged from 2000 to 2005. No logging is planned in the near future. In the agricultural zone, 64% of plant production comprises perennial crops (473 ha of hay) and 36% are annual crops, including 103 ha of cereals, 68 ha of corn, 11 ha of horticultural crops, and 2 ha of berries.

2.2. Sources of hydrological and photographic data

In the absence of flow gauging station on the Matambin River, the closest station (station 052601) is located on the Maskinongé River, at the Canadian National (CN) railway bridge, near the town of Sainte-Ursule (1030 km^2), where flows have been measured daily since 1925. The station is located 20 km from the confluence of the Matambin and Maskinongé Rivers. Because no flow data for this station are available for the period from 1973 to 1978, flow data measured at the Joliette station in the L'Assomption River watershed, to the southeast of the Matambin River watershed, were also analyzed. In Joliette, the L'Assomption River watershed has a surface area of 1390 km^2, which is similar to that of the Maskinongé River watershed at the Sainte-Ursule station. Moreover, both rivers originate in the Canadian Shield and flow through the St Lawrence Lowlands before flowing out into the St Lawrence River. Consequently, their watersheds have similar climatic and physiographic characteristics, as shown by coefficients of correlation derived between flows in the two rivers (**Table 1**). In Joliette, flows have been measured since 1922, 3 years prior to the start of measurements on the Maskinongé River. Historical daily and monthly flow data for the two rivers (Maskinongé and L'Assomption) were provided by the Centre d'Expertise Hydrique du Québec (http://www.cehq.gouv.qc.ca/, viewed 2013-05-09).

Flows	Winter	Spring	Summer	Fall
Daily maximum flows	0.8233	0.5815	0.8327	0.9191
Mean daily flows	0.8012	0.7477	0.8483	0.9616

Note: All coefficients of correlation are statistically significant at the 1% level.

Table 1. Coefficients of correlation calculated between flows measured at the Joliette (L'Assomption River) and Sainte-Ursule (Maskinongé River) stations during the period from 1930 to 2010.

Air photographs were obtained for several decades from 1930 to 2008. However, for this study, air photos taken in the 1940s were excluded because they were taken during a wartime period. Photos from the 1950s were also excluded because of their inadequate scale (errors on the outline of banks would have been too large). Results for photos taken in the 1980s are not presented because they are not significantly different (no significant morphological change) from those for the following decade. Although we attempted to select air photographs taken at similar time and/or under similar hydrological conditions, it was sometimes difficult to do so given the photos available. However, given that the outlines of banks were constrained at bankfull flow level, this is not an issue if photos were not taken when flows were above bankfull flows. **Table 2** shows some of the characteristics of selected air photographs. Because of the low coverage (two photos covering only the lower course), photographs taken in 1975 were only used for assessing land use in the agricultural zone.

No data exist on soil erosion in the watershed. As for measurements of suspended solids carried by the Matambin River and other Quebec rivers, they are only available for the period from 2013 to 2015, and were taken by the Quebec Ministère du Développement durable, de l'Environnement et de la Lutte contre les changements climatiques (http://www.mddelcc.gouv.qc.ca/eau/Atlas_interactif/stations/stations_rivieres.asp, viewed 2017-02-09).

2.3. Data analysis

2.3.1. Analysis of hydrologic data

As part of this study is aimed at explaining the potential influence of streamflow on the morphological evolution of the Matambin River channel, two types of seasonal flow series were analyzed: a series of seasonal maximum daily flows and a series of seasonal mean daily flows. The first series comprises the highest daily flows measured each season and each year. The second series comprises the mean values of daily flows in each season and each year. Years were divided in seasons following an approach used by many authors in Quebec [19–21] for ease of comparison of results. Thus, seasons were defined as: winter (January–March), spring (April–June), summer (July–September), and fall (October–December). This is the scheme that best reflects the natural hydrological cycle in Quebec, Canada.

Date taken	Series	Scale	Daily discharge (m^3/s)*
08 October 1935	A5259	1/15,000	8.5
02 September 1964	Q64545	1/15,840	2.38
30 June 1975	Q75330	1/15,000	-
23 July 1997	HMQ97138	1/15,000	4.83
12 June 2008	Q08001	1/15,000	17.54

*Discharge measured at the Sainte-Ursule station on the Maskinongé River. These flows are lower than bankfull flow. Photographs for 1964 were enlarged to 1:15,000.

Table 2. Characteristics of analyzed air photographs.

The two hydrologic series were analyzed to constrain the temporal variability of flow. Because the first air photographs available date back to the 1930s, the same year was taken as the beginning of the hydrologic analysis. It is worth recalling that many statistical tests and methods can be used to analyze the temporal variability of streamflow [22]. However, the two approaches widely used in hydrology are the Mann-Kendall test and the linear regression method. These two tests are not without drawbacks, however:

- They do not detect a gradual or abrupt change in mean or variance of hydrologic series analyzed.
- They do not constrain the precise timing of such changes.

The Lombard method [23] was used to overcome these two weaknesses. This is a general method for detecting a gradual or abrupt change in the mean value of a hydrological series, given that methods commonly used in hydrology (e.g., Petittt method) do not allow the detection of gradual and abrupt changes. From a statistical standpoint, the mathematical aspects of this method are described in Refs. [23, 24]. Applications of the Lombard method in hydrology and geomorphology are presented in many papers (e.g., [24–26]). It should be mentioned that it was possible to use the Lombard method in this study because the series analyzed did not show any autocorrelation.

Given a series of observations, noted $X_1, \ldots, X_n$, where X_i is the observation taken at time $T = i$. These observations are supposed to be independent. If μ_i refers to the theoretical mean of X_i, then a possible pattern for the mean is given by Lombard's smooth-change model where

$$\mu_i = \begin{cases} \theta_1 & \text{if } 1 \le i \le T_1 \\ \theta_1 + \dfrac{(i - T_1)(\theta_2 - \theta_1)}{T_2 - T_1} & \text{if } T_1 < i \le T_2 \\ \theta_2 & \text{if } T_2 < i \le n \end{cases} \tag{1}$$

In other words, the mean changes gradually from θ_1 to θ_2 between times T_1 and T_2. As a special case, one has the usual abrupt-change model when $T_2 = T_1 + 1$.

In order to test formally whether the mean in a series is stable or rather follows model (1), one can use the statistical procedure introduced by Lombard [23]. To this end, define R_i as the rank of X_i among $X_1, \ldots, X_n$. Introduce the Wilcoxon score function $\phi(u) = 2u - 1$ and define the rank score of X_i by

$$Z_i = \frac{1}{\sigma_\phi}\left\{\phi\left(\frac{R_i}{n+1}\right) - \overline{\phi}\right\}, \quad i \in \{1, \ldots, n\} \tag{2}$$

where

$$\phi = \frac{1}{n}\sum_{i=1}^{n}\phi\left(\frac{i}{n+1}\right) \quad \text{and} \quad \sigma_\phi^2 = \frac{1}{n}\sum_{i=1}^{n}\left\{\phi\frac{i}{n+1} - \overline{\phi}\right\}^2 \tag{3}$$

Lombard's test statistic is

$$S_n = \frac{1}{n^5}\sum_{T_1=1}^{n-1}\sum_{T_2=T_1+1}^{n} L_{T_1T_2}^2 \tag{4}$$

where

$$L_{T_1,T_2} = \sum_{j=T_1+1}^{T_2}\sum_{i=1}^{j} Z_i \tag{5}$$

At the 95% confidence interval, one concludes that the mean of the series changes significantly according to a pattern of type (1) whenever $S_n > 0.0403$. This value corresponds to Test Lombard statistic value (see [23]) defining the significance level at 5% for the test. Note that the equation proposed by [23] to detect multiple abrupt changes in the mean of a statistical series was also applied. This formula confirmed results obtained using Eq. (1).

2.3.2. *Analysis of air photographs*

We have described the method used in detail in some of our previous work [1–3]. Diachronic analysis of aerial photographs taken at different times was used to obtain data on the evolution of forested surface and on morphology (**Table 2**). Aerial photograph analysis was done in three steps: georeferencing, orthorectification, and mosaicing. Before being georeferenced, photos were first scanned at a resolution of 600 dots per inch (DPI) and 8 bits (color depth). The resulting resolution is estimated at 0.6 m (for photos taken at 1:15,000). The root mean square error (RMSE) of the geometric rectification on photos captured at different times was <0.1 m after the scanned photos underwent a second-order polynomial transformation. The Geomatica (PCI Geomatics 2003) OrthoEngine package was used for georeferencing, which involved the use of 10 ground control points spatially distributed over the whole photo area. UTM coordinates were taken from a 1:15,000 topographic map (surveyed 1995) from the Base de données topographiques du Québec (Quebec Topographic Database). Photos were orthorectified after being georeferenced, which consisted in correcting for spatial distortion arising from the inclination of the camera and/or terrain relief. Camera orientation parameters, the various known reference points, and the digital elevation model (DEM) were used for this correction. Mosaics of the georeferenced and orthorectified photos were then assembled (one mosaic for each year photos were taken) to produce an orthophotograph covering the entire study area.

After aerial photograph processing, the ESRI ArcGIS software was used to digitize river banks (at bankfull level as defined by the vegetation edge). A three-dimensional digital stereoscope (forward overlapping aerial photo pairs), which allows three-dimensional viewing of the channel, was used to limit error on bank outline. Automated drawing of straight lines perpendicular to the channel using an ArcGIS extension developed at the Laboratoire Interdisciplinaire d'Application en Géomatique Environnementale (LIAGE) and integrated into ArcGIS was used to measure channel width. This step eliminates human error associated with drawing of perpendicular lines and thus in channel bankfull width measurements. Automated drawing of perpendicular lines at the same location between two points separated by the same distance on air photos taken at different times, which is difficult to do manually, is also possible using this software. After banks are delineated, the perpendicular lines are drawn by first finding the midpoint between the two banks in each cross-section, then joining all midpoints into a centerline, and finally drawing lines perpendicular to this centerline at all midpoints. All air photos taken at different times were processed in this way. More than 20 bankfull channel width measurements (depending on the total length of the analyzed reach) were taken with the software at a 100-m regular spacing. This interval was selected to reduce the spatial autocorrelation effect and make it possible to analyze long statistical series. Channel sides delineated from aerial photography were validated with ground observations. Fairly precise definition of channel side limits used to calculate the channel bankfull width was possible using this technique. The validity of bank delineation based on aerial photographs was tested using field measurements taken during several field visits. The fact that flows at the time the photographs were taken were less than bankfull made it possible to carry out diachronic analysis of the aerial photographs (see **Table 2**).

Air photo interpretation (width measurement from aerial photos) was done by two independent operators to determine the maximum error on bankfull width measurements associated with photo interpretation. The maximum difference in bankfull mean width at a given station between the values obtained by the two operators was <1 m, and this value is deemed to represent the maximum error on channel width measurements from photo interpretation at a given station. The same results were obtained from analysis of the temporal variability of channel width carried out by both the operators. The non-parametric Kruskal-Wallis and parametric analysis of variance (ANOVA) tests were used to compare bankfull width values measured at different times for each river.

The outlines of forested settings were traced from available digital land use maps using the ArcGIS software (version 9.2) to delineate agricultural settings and assess the evolution of agricultural and forested surface areas over time. Reforested area boundaries on photographs taken at different times were superimposed within the study framework.

3. Results

3.1. Temporal variability of flows

Values of the Lombard method S_n statistic are presented in **Table 3**. Only maximum daily flows in winter changed significantly for the two rivers. However, for the Maskinongé River, winter mean daily flows also increased significantly. This increase, which is abrupt for both rivers, occurred in the early 1970s. No significant change is observed for the other three seasons. However, it is important to note that, despite this lack of change in mean value of the magnitude of maximum flows, a higher frequency of high magnitude flows is observed for the L'Assomption River after 1970 (**Table 4**) for all four seasons, particularly in springtime. A comparison of the interannual variability of maximum daily flows for each of the four seasons for both rivers is presented in **Figures 3–6**.

3.2. Evolution of forest cover in the agricultural zone within the Matambin River watershed

In 2008, 81.56 km^2 of the 98.03 km^2 of the Matambin River watershed were forested or 83.2% of the watershed, the deforested area thus representing 16.47 km^2 or 16.8% of the watershed. Diachronic analysis of air photographs of forested areas revealed that, in the agricultural zone, the proportion of forests increased from 1935 to 2008, from 13.2% (1.27 km^2) in 1935 to 27.1% (2.12 km^2) in 2008. Hence, this proportion more than doubled over the years at the expense

Seasons	Maskinongé River			L'Assomption River		
Seasonal daily maximum flows						
	Sn	T1	T2	Sn	T1	T2
Winter	**0.0608**	1970	1971	**0.0827**	1971	1972
Spring	0.0119	–	–	0.0249	–	–
Summer	0.0302	–	–	0.0058	–	–
Fall	0.0400	–	–	0.0337	–	–
Seasonal mean daily flows						
Winter	0.0525	1970	1971	0.0024	–	–
Spring	0.0345	–	–	0.0014	–	–
Summer	0.0149	–	–	0.0339	–	–
Fall	0.0353	–	–	0.0330	–	–

Note: Lombard test results. Values of S_n (Lombard test statistics) >0.0403 that are statistically significant at the 5% level are shown in bold. T_1 and T_2 are the years of start and end, respectively, of the shift in a mean value.

Table 3. Analysis of the interannual variability of flows in the Maskinongé and L'Assomption Rivers during the period from 1930 to 2010.

Seasons	1935–1964	1965–1994
Winter (≥100 m^3/s)	2	4
Spring (≥250 m^3/s)	2	5
Summer (≥100 m^3/s)	0	2
Fall (≥150 m^3/s)	1	2

Table 4. Comparison of the frequency of strong seasonal maximum daily flows for the L'Assomption River at the Joliette station over the period from 1930 to 2010.

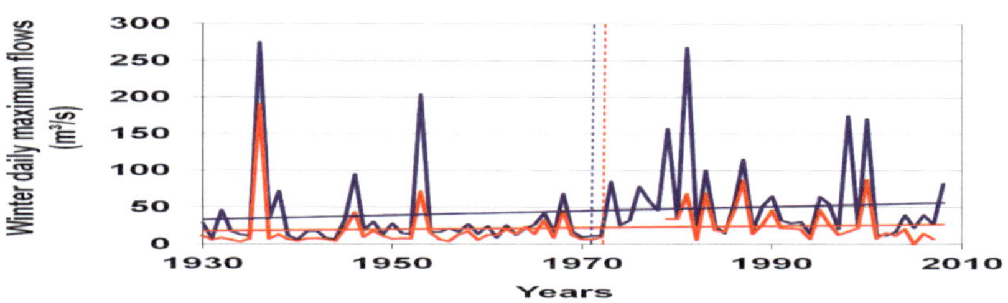

Figure 3. Variability of winter daily maximum flows at the Joliette station on the L'Assomption River (blue curve or dark grey) and Sainte-Ursule station on the Maskinongé River (red curve or fair grey) during the period from 1930 to 2010. Vertical dotted lines represent years of shifts in mean values of flows (blue curve or dark grey) and (red curve or fair grey).

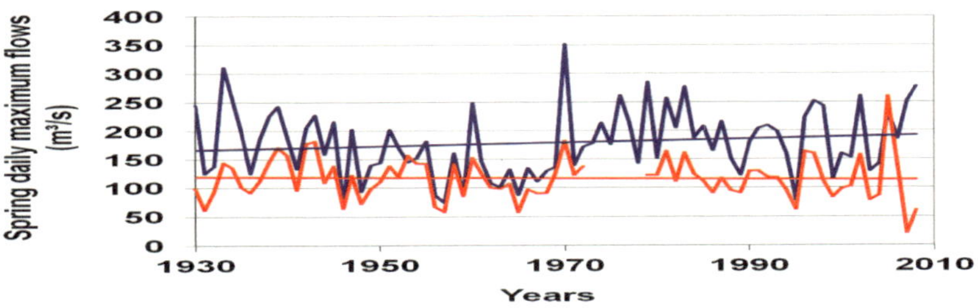

Figure 4. Variability of spring daily maximum flows at the Joliette station on the L'Assomption River (blue curve or dark grey) and Sainte-Ursule station on the Maskinongé River (red curve or fair grey) during the period from 1930 to 2010.

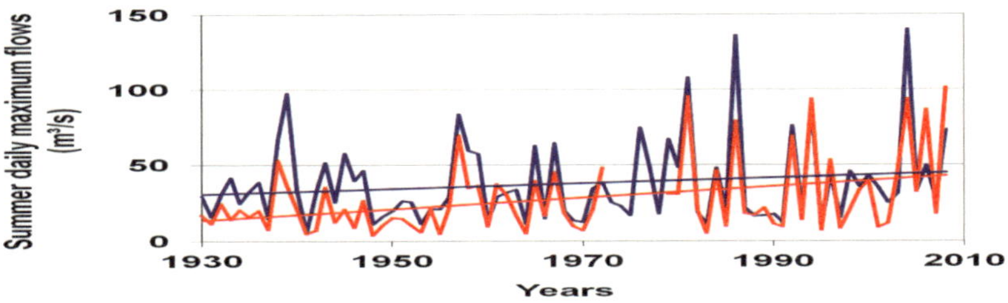

Figure 5. Variability of summer daily maximum flows at the Joliette station on the L'Assomption River (blue curve or dark grey) and Sainte-Ursule station on the Maskinongé River (red curve or fair grey) during the period from 1930 to 2010.

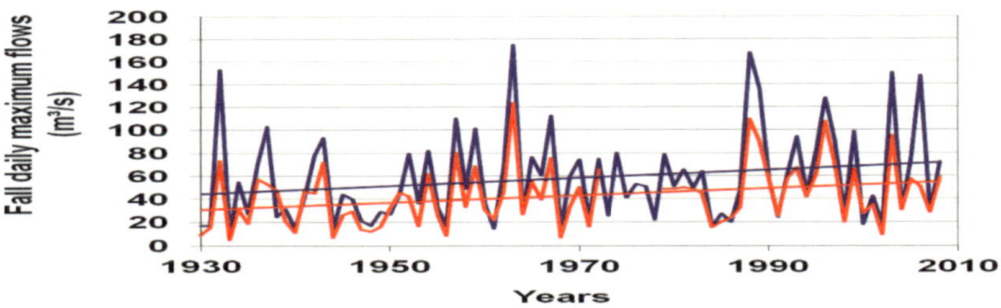

Figure 6. Variability of fall daily maximum flows at the Joliette station on the L'Assomption River (blue curve or dark grey) and Sainte-Ursule station on the Maskinongé River (red curve or fair grey) during the period from 1930 to 2010.

of cultivated land. While forested land increased gradually through the years, two increases took place between 1935 and 1964 (+32.4%) and between 1975 and 1997 (+23.92%) (**Table 5** and **Figure 7**). In terms of cumulative percentage, the reforested surface area went from 32.37% in 1964 to 56.27% in 1997 in the agricultural zone. It is worth pointing out that reforestation activities began only in the 1950s throughout Quebec, as in many developed countries, as a result of the decline of farming populations for various reasons: intensification (motorization, mechanization, use of fertilizers, selection of high-yield seeds, use of pesticides, etc.) and increase in secondary and tertiary industry activities (e.g., [4–5, 27]). Thus, the significant 32.4% increase in forested surface area in the watershed took place between 1950 and 1964.

	1935	1964	1975	1997	2008
Forested surface area	1.27	1.68	1.71	2.08	2.12
Increase rate (%)	–	32.3	1.7	22.2	1.6
Cumulated increase rate (%)	–	32.3	34	56.2	57.8

Table 5. Evolution of forested surface area (reforestation in km^2) in the agricultural zone in the Matambin River watershed (1935–2008).

3.3. Evolution of Matambin River channel bankfull width and sinuosity

Analysis of the longitudinal variability of the bankfull width of the Matambin River channel using the Lombard method revealed no significant change in the mean value of this width (results not presented) from upstream to downstream despite the increase in watershed surface area. This is due to the fact that the river has very few tributaries in the agricultural area (St Lawrence Lowlands) through which its lower reaches flow. As far as temporal variability is concerned, mean values of bankfull width and sinuosity are presented in **Table 6** and **Figure 8**. Results of the comparison of these mean values using the Kruskal-Wallis method and the Tukey test are shown in **Tables 7** and **8**, respectively. These tables show that mean bankfull width first decreased from 1935 to 1964 (−21%) and then increased gradually (+12%) from 1964 to 2008. The Kruskal-Wallis test reveals that the decrease is statistically significant at the 1% level. The Tukey test (**Table 8**) shows that the mean value of channel bankfull width in 1935 is significantly different from values for subsequent years. In contrast, the increase in width observed after 1964 is not statistically significant because mean values of width measured for 1964, 1997, and 2008 are not significantly different. Sinuosity, for its part, varied little over time, increasing very slightly between 1934 and 1964, and then decreasing after 1964 due to the cutoff of some meanders (**Figure 9**).

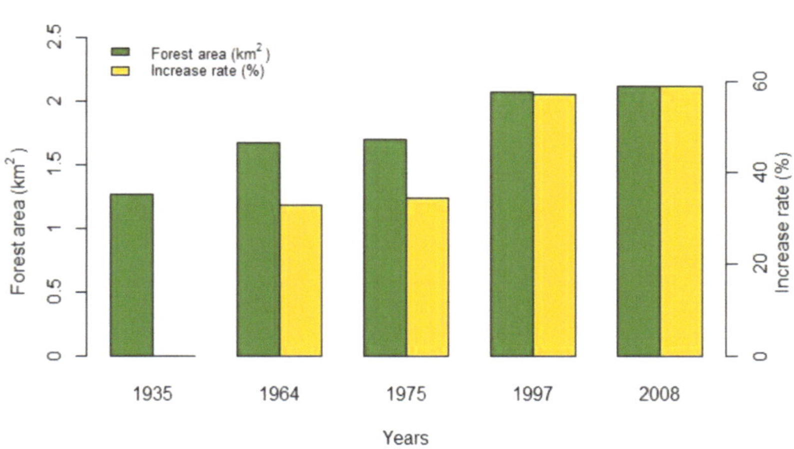

Figure 7. Evolution of the forested surface area in the agricultural zone. Green bars = forested surface area in km^2; yellow bars = cumulative percentage of agricultural surface area.

Year	Mean	Standard deviation	Sinuosity
1935	16.5 (935)	5.45	1.77
1964	13 (806)*	4.63	1.51
1997	13.4 (935)	4.52	1.75
2008	14.3 (935)	4.38	1.71

Note: ( ) = total number of bankfull width measurements.

*Air photos lacking for part of the channel.

Table 6. Comparison of the temporal evolution of the mean value of bankfull width (*m*) and the sinuosity of the Matambin River channel in the reforested agricultural zone over the period from 1935 to 2008.

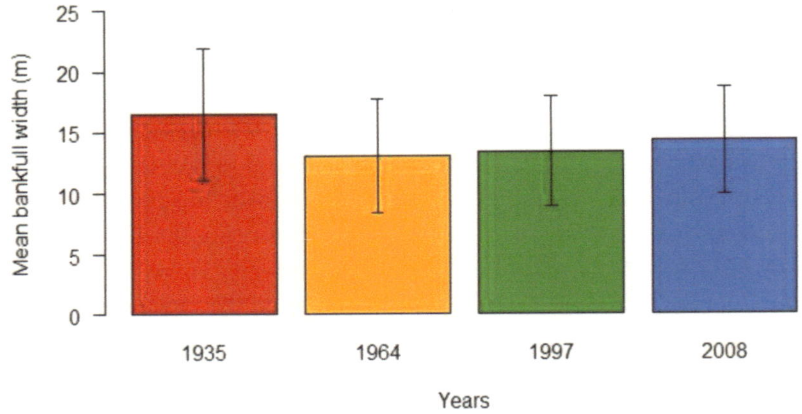

Figure 8. Evolution of mean bankfull width of the Matambin River. Vertical lines represent standard deviation values for the means.

Sources	SM	DL	CM	F	p-Value
Years	6478.771	3	2159.590	94.956	0.000
Error	82307.529	3619	22.743		

Note: SM = sum of squares; DL= number of degrees of freedom; F = calculated value of the Fisher-Snedecor test.

Table 7. Comparison of mean values of bankfull width of the Matambin River channel in the agricultural zone using analysis of variance with a single classification criterion.

Years	1935	1964	1997	2008
1935	1			
1964	**0.000**	1		
1997	**0.000**	0.217	1	
2008	**0.000**	**0.001**	**0.000**	1

Note: Probability values < 0.05 are statistically significant at the 5% level.

Table 8. Comparison of mean values of bankfull width of the Matambin River channel in the agricultural zone using the *post hoc* Tukey test.

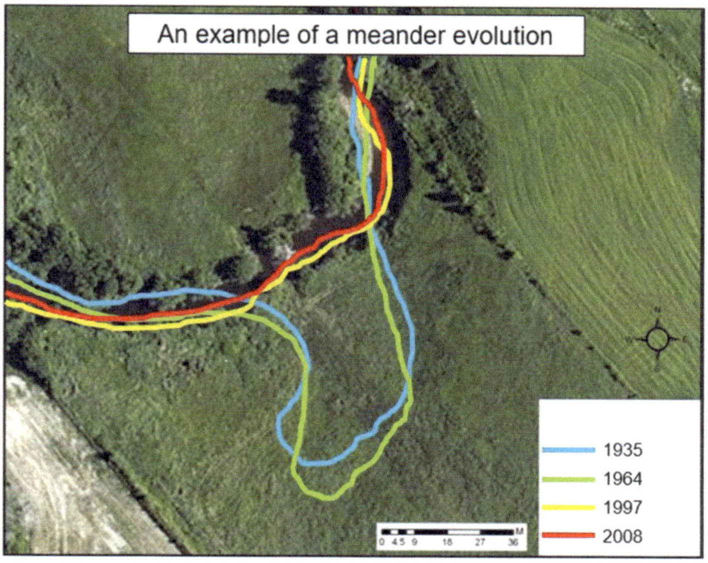

Figure 9. Example of meander cutoff. Lines represent the central axis of the river channel in different years.

4. Discussion and conclusion

In many developed countries, a significant decrease in farming populations has been observed since the second half of the last century. This decrease has led to cultivated land being abandoned, and this land has been taken over by natural vegetation and/or reforested by humans. The hydrological and morphological consequences of these land-use changes have been studied in several countries (e.g., [4, 6, 8–9, 11, 14, 16, 28]). However, despite the increase in reforested land since the 1950s in the St Lawrence Lowlands, no study has yet looked at the morphological impacts of such changes in Quebec. This study analyzed the morphological evolution of the Matambin River channel, along with the reforestation of previously cultivated land. Diachronic analysis of air photographs revealed that the reforested surface area has more than doubled in the agricultural zone from 1950 to 2008. During the same period, two phases of morphological evolution of channel width are observed: a first phase characterized by a significant decrease (−21%) in mean bankfull width which occurred between 1935 and 1964, followed by a second phase characterized by a trend of increasing width from 1964 to 2008. There were no concomitant changes in sinuosity, which changed very little over time.

Prior to the start of reforestation in the agricultural zone during the 1950s, agricultural activities generated large amounts of suspended sediments in this zone caused by soil erosion. Rivers that drain these lands transport much larger suspended loads than rivers draining forested areas, as **Figure 10** clearly shows. The deposition of these large amounts of suspended sediments in the alluvial plain and along the banks could account for the decrease in mean width observed from 1935 to 1964, taking into account the fact that reforestation started only in the 1950s. In addition, *the decrease in* frequency of strong floods observed (see **Table 4**) in the area over the same period would have promoted this deposition while reducing the capacity of the river to carry away all sediments produced by soil erosion (e.g., [29]).

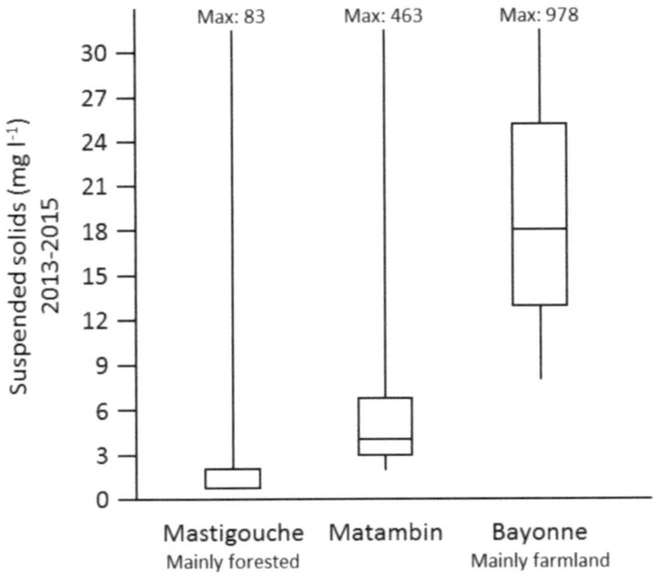

Figure 10. Comparison of mean and maximum suspended solid concentrations carried by the Mastigouche (forested watershed, drainage area: 589.4 km^2), Matambin (reforested agricultural watershed, drainage area: 95.5 km^2), and Bayonne (agricultural watershed, drainage area: 363.3 km^2) Rivers from 2013 to 2015, in Quebec.

After 1964, channel width tended to increase gradually, and this increase, although slow, seems to be ongoing. Two factors may account for this change: (1) the increased frequency of strong flood flows in the area (see **Table 4**), and (2) the significant decrease in the amount of sediments derived from erosion following reforestation. As far as this latter factor is concerned, no long-term data are available on suspended loads carried by the Matambin River. However, a comparison of mean and maximum suspended solid concentrations over the period from 2013 to 2015 in three Quebec watersheds characterized by different land uses clearly shows that the Bayonne River (agricultural watershed) carries on average at least three times the suspended load as the Matambin (reforested agricultural watershed) and Mastigouche (mainly forested watershed) Rivers (**Figure 10**). From these results, we can state with certainty that the Matambin River also carried large suspended loads prior to reforestation of its watershed because it drains the same type of soils as the Bayonne River. Be that as it may, [7] observed a similar trend in the Warche River watershed upstream from the Butgenbach reservoir in Belgium. After farming ceased, giving way to prairies and reforestation by humans, the Warche River channel grew significantly over time due to the decreasing amount of sediments derived from formerly cultivated soils. Moreover, as in the case of the Matambin River, this channel widening was accompanied by very slight changes in channel sinuosity. In the case of the Matambin River, the effect of lower suspended loads on channel widening is thought to be sustained by the increasing frequency of strong flood flows. However, unlike the Warche River, the Matambin River channel is sometimes obstructed by dead wood in many places. The presence of such obstacles may reduce the erosional capacity of the river (e.g., [28]), thus slowing down the channel widening process.

However, in a number of European countries, several studies have shown that reforestation of old agricultural zones has led to a significant decrease in the width of channels. These studies, however, were done in areas characterized by rugged topography (mountainous areas or plateaus) and in rivers with bedloads consisting primarily of gravel or boulders. In addition, in most of these studied watersheds, streamflow decreased significantly due to a decrease in the amount of rain. Even in Quebec, in this type of rivers, the decrease in suspended load observed downstream from some dams has led to a decrease in channel width in boulder reaches (e.g., [30]). This decrease is also accounted for by the near disappearance of all floods with >2 years recurrence intervals after construction of the dams. In contrast, in sandy reaches consisting primarily of easily erodible fines, no change in channel width has been observed as a result of the decrease in strong flood flows.

Finally, one of the goals of this study was to check the validity of relationships set forth by [31] to account for changes in morphological variables of channels as a function of liquid and solid flows (bedload and suspended load). As far as width is concerned, according to these relationships, a decrease in streamflows associated with an increase in solid flows could lead to an increase or a decrease in width. In the case of the Matambin River, a significant decrease in width is observed between 1935 and 1964, a period during which the frequency of strong flood flows decreased while the amount of suspended sediments remained relatively large. In contrast, from 1964 to 2008, these changes in liquid and solid flows were inverted in the watershed, resulting in a trend of increasing channel width. However, unlike width, the sinuosity of the Matambin River was not very strongly affected by changes in the amount of liquid and solid flows in the watershed. Given the above, from a management standpoint, reforestation of old farmlands primarily leads to the cessation of channel narrowing and has little effect on channel sinuosity, allowing the conservation of most available habitats for aquatic and semi-aquatic organisms.

Author details

Ali Assani*, Lisanne Chauvette and Stéphane Campeau

*Address all correspondence to: Ali.Assani@uqtr.ca

Department of Environmental Sciences, University of Quebec at Trois-Rivières, Trois-Rivières, Québec, Canada

References

[1] Alibert M, Assani AA, Gratton D, Leroux D, Laurencelle M. Statistical analysis of the evolution of semialluvial stream channel upstream from an inversion-type reservoir: The case of the Matawin River (Quebec, Canada). Geomorphology. 2011;**131**:28-34

[2] Aubry L, Assani AA, Biron S, Gratton D. Comparison of the hydromorphological evolution of the L'Assomption and Oureau River channels (Québec, Canada). River Research and Applications. 2012;**29**:979-990

[3] Vadnais ME, Assani AA, Landry R, Leroux D, Gratton D. Analysis of the effects of human activities on the hydromorphological evolution channel of the Saint-Maurice River downstream from La Gabelle dam (Quebec, Canada). Geomorphology. 2012;**175-176**:199-208

[4] Keesstra SD, van Huissteden J, Vandenberghe J, Van Dam O, de Gier J, Pleizier ID. Evolution of the morphology of the river Dragonja (SW Slovenia) due to land-use changes. Geomorphology. 2005;**69**:191-207

[5] Liébault F, Piégay H. Causes of 20th century channel narrowing in mountain and piedmont rivers of southeastern France. Earth Surfaces Processes and Landforms. 2002;**27**:425-444

[6] Stott T. A comparison of stream bank eroison processes on the forested and moorland streams in the Balquhidder catchment, Central Scotland. Earth Surfaces Processes and Landforms. 1997;**22**:383-399

[7] Assani AA, Petit F, Buffin-Bélanger T, Roy AG. Analyse de la variation spatio-temporelle de la morphologie du chenal de la Warche en amont du barrage de Butgenbach (Belgique). Zeitschrift fur Geomorphologie. 2003;**47**:469-483

[8] Cadol D, Rathburn SI, Cooper DJ. Aerial photographic analysis of channel narrowing and vegetation expansion in Canyon De Chely national monument, Arizona, USA. 1935-2004. River Research and Applications. 2011;**27**:841-856

[9] Kasai M, Brierley G J, Page MJ, Marutani T, Trustrum NA. Impacts of land use change on patterns of sediment flux in Weraamaia catchment, New Zealand. Catena. 2005;**64**:27-60

[10] Lach J, Wyzga B. Channel incision and flow increase of the upper Wisloka river, Southern Poland, subsequent to the reafforestation of its catchment. Earth Surfaces Processes and Landforms. 2002;**27**:445-462

[11] Mount NJ, Smith GHS, Stott TA. An assessment of the impact afforestation on lowland river reaches: The Afon Tranon, mid-Wales. Geomorphology. 2005;**64**:255-269

[12] Mueller EN, Francke T, Batalle RJ, Bronsert A. Modelling the effects of land-use change on runoff and sediment yield for a meso-scale catchment in the Southern Pyrenees. Catena. 2009;**79**:288-296

[13] Murgatroyd AL, Ternan JL. The impact of afforestation on stream bank erosion and channel form. Earth Surfaces Processes and Landforms. 1983;**8**:357-369

[14] Porto P, Walling DE, Callegari G. Investigation the effects of afforestation on soil erosion and sediment mobilisation in two small catchments in Southern Italy. Catena. 2009;**79**:181-188

[15] Stott T. Forestry effects on bedload yields in the mountain streams. Applied Geography. 1997b;**17**:55-78

[16] Van Rompaey AJJ, Govers G, Puttemans C. Modelling land use changes and their impact on soil erosion and sediment supply to rivers. Earth Surfaces Processes and Landforms. 2002;**27**:481-494

[17] Farley KA, Jobbágy EG, Jackson RB. Effects of afforestation on water yield: A global synthesis with implications for policy. Global Change Biology. 2005;**11**:1565-1576

[18] Association pour la gestion intégrée de la rivière Maskinongé (AGIR). Portrait et diagnostic du Plan directeur de l'eau du bassin versant de la rivière Maskinongé, Maskinongé (Québec). Technical Report; 2008

[19] Assani A A, Gravel E, Buffin-Bélanger T, Roy AG. Impacts des barrages sur les débits annuels minimums en fonction des régimes hydrologiques artificialisés au Québec (Canada). Revue Des Sciences De L'eau. 2005;**18**:103-127

[20] Lajoie F, Assani AA, Roy AG, Mesfioui M. Impacts of monthly flow characteristics. The influence of watershed size and seasons. Journal of Hydrology. 207;**334**:423-439

[21] Matteau M, Assani AA, Mesfioui M. Application of multivariate statistical analysis methods to the dam hydrologic impact studies. Journal of Hydrology. 2009;**371**:120-128

[22] Kundzewicz ZW, Graczyk D, Maureer T, Pinskwar I, Radziejewski M, Svensson C, Szwed M. Trend detection in river flow series: 1. Annual maximum flow. Hydrological Sciences Journal. 2005;**50**:797-810

[23] Lombard F. Rank tests for changepoint problems. Biometrika. 1987;**74**:615-624

[24] Quessy JF, Favre AC, Saïd M, Champagne M. Statistical inference in Lombard's smooth-change model. Environmetrics. 2011;**22**:882-893

[25] Assani AA, Chalifour A, Légaré G, Manouane C-S, Leroux D. Temporal regionalization of 7-day low flows in The St-Lawrence watershed in Quebec (Canada). Water Resources Management. 2011b;**25**:3559-3574

[26] Landry R, Assani AA, Biron S, Quessy JF. The management modes of seasonal floods and their impacts on the relationship between climate and streamflow downstream from dams in Quebec (Canada). Rivers Research and Applications. 2014;**30**:287-298

[27] Stott T, Mount N. Plantation forestry impacts on sediment yields and downstream channel dynamics in the UK: A review. Progress in Physical Geography. 2004;**28**:197-240

[28] Assani AA, Petit F. Log-jam effects on bed-load mobility from experiments conducted in a small gravel-bed forest ditch. Catena. 1995;**25**:117-126

[29] Kuhnle RA, Binger RI, Foster GR, Grissinger EH. Land use changes and sediment transport in Goodwin Crekk. In: Klingeman PC, Beschta, RL, Komar, PD, Bradley, JB, editors. Gravel Bed Rivers in the Environment. Colorado: Water Resources Publications; 1998. pp. 279-291

[30] Hubert K. Effets des changements des caractéristiques de crues sur l'évolution morphologique de la rivière Matawin en aval du réservoir Taureau (Québec, Canada). Trois-Rivières: Université du Québec à Trois-Rivières; 2011.

[31] Schumm SA. River metamorphosis. Journal of Hydraulics Division. Proceedings of the America Society of Civil Engineers. 1969;**HY1**:255-273

Chapter 4

The Modelling of Coastal Cliffs and Future Trends

Ricardo Castedo, Carlos Paredes,
Rogelio de la Vega-Panizo and Anastasio P. Santos

Additional information is available at the end of the chapter

http://dx.doi.org/10.5772/intechopen.68445

Abstract

About 80% of the world's oceanic shorelines include diverse types of cliffed and rocky coasts: plunging cliffs, bluffs backing beaches and rocky shore platforms. In combination, approximately 60% of the world's population lives within 60 km of the coast. Rapidly retreating soft cliffs may be found worldwide and are particularly vulnerable to changes in the forcing factors. The study and analysis of the rate of change in shoreline position through time is important or even imperative for coastal management. The development of cliff erosion predictive models is mainly limited to geomorphological data because of the complex interactions between physical-chemical processes acting simultaneously in time and space that result in large scale variations. Current historical extrapolation models use historical recession data, but different environments with the same historical values can produce identical annual retreat characteristics despite the potential responses to a changing environment being unequal. For that reason, process-response models (PRMs) are necessary to provide quantitative predictions of the effects of natural and human-induced changes that cannot be predicted using other models. Several models are explained and discussed, including a process-response model, based on real data at Holderness Coast (UK).

Keywords: coastal recession model, clay cliffs, Holderness Coast, numerical model, climate change

1. Introduction

Approximately 60% of the world's population lives within a band 60 km wide from the coastline towards coastal hinterland: the coastal zone. The "coastal zone" is defined in a World Bank publication as "*the interface where the land meets the ocean, encompassing shoreline environments as well as adjacent coastal waters. Its components can include river deltas, coastal plains, wetlands, beaches and dunes, reefs, mangrove forests, lagoons and other coastal features*". The increasing world

INTECH
open science | open minds

© 2017 The Author(s). Licensee InTech. This chapter is distributed under the terms of the Creative Commons Attribution License (http://creativecommons.org/licenses/by/3.0), which permits unrestricted use, distribution, and reproduction in any medium, provided the original work is properly cited.

population, from 6 billion in 2000 to 9–10 billion by 2100, will lead to accelerating occupation of coastal zones even if these areas are seriously threatened by erosive processes [1–3], if so, an integrated management of resources and spaces is urgently required. Coastal recession and instabilities increases the risk on properties, people and their economic activities. Because of these and to improve coastal management and engineering intervention, it is critical to understand the coastal change pattern. The shoreline is dynamic in nature with different phenomena that acts at different temporal and spatial scales [4]. This area is also very sensible to changes in the conditions like sea-level rise (SLR) or extreme weather events, which are mostly related to the acceleration of coastal retreat [5–7]. Some rapidly retreating areas like the Great Lakes of North America, England (north-eastern coasts), and parts of the United States (north-eastern coasts) are highly exposed to changes in these factors [8–12]. On rapidly retreating coasts, it is not just the geographic position of the cliff face that is important, but the entire interaction of the land with the constantly changing hydrodynamic regime creating a complex coastal system [4, 7, 13, 14]. Therefore, coastal managers and engineers must also consider the on- and offshore environment; weather; climate; the geotechnical properties of the slope forming materials; the influence of man-made structures and also the effects of climate change and the associated sea level rise. This means that many factors control the complex behaviour and plan or profile shape of the coastline.

As well as assisting with the prediction of social implications of land loss, increasing the understanding of cliff erosion is of great importance in the wider science that underpins coastal engineering. However, in order to improve different coastal problems, the critical data are the shoreline position through time and/or its rate of change. These problems can be hazard zoning, developing appropriate management responses and coastal zone governance, assessing the possible effects of sea-level rise and conceptual or predictive modelling of coastal morphodynamics [15–19]. Hence, appropriate and practical tools are needed to help forecast future shoreline evolution, especially where the shoreline responds in a non-linear fashion due to non-uniform and non-homogeneous variations in the geology, the environment, the hydrodynamic regime and the changing climate. The tools, in terms of computational models, that are designed should be able to calculate an accurate approach of the coastal cliff state (position and shape) on the meso-timescale (1–100 years), although longer term forecast may be useful or required for more long-lasting structures. Although predictions have been based on historical rates of recession and on a variety of empirical and probabilistic methods [13, 18, 20], historical rates of change cannot be assumed to continue into the future, especially under changing conditions like geology, environment, hydrodynamic regime and climate [7]. Process-response models (PRMs) [21–23] are needed to address these issues and provide quantitative predictions of the effects of natural and human-induced changes, which cannot be predicted from statistical analysis of historic recession data or other models such as the Bruun-rule model. For example, Bray and Hook [13] based on empirical methods like the "modified Bruun" model (also called historical trend analysis—[24]) related the erosion only with the sea-level rise. The application of the latter model provides deterministic predictions of recession, without truly reflecting the uncertainty and process variability.

There are few published reliable process-response models for cliff recession. These are based on functional relationships between the dominant physical processes covering the shore-face, beach and cliff [21–23]. However, additional models are needed to simulate basal erosion and the resultant effects transmitted by gravitational movements up the cliff to the backscar, and this forms a major research gap [13], although progress is possible by adapting geotechnical stability analyses. Ashton et al. [25] developed a model called CEM (Coastline Evolution Model) focused on non-cliffed coasts that includes a simplified geology, wave directions and sediment currents, and simply human activities. Walkden and Hall [22, 23] designed the first process-response model (PRM) that includes individual modules to represent the evolution of soft rock-shore profiles. These modules are, for example, rock erosion, beach protection, near shore wave transformation or sea level changes. In the model called SCAPE, at every time step, marine conditions are read from input files, and the new profile determines the system state. This whole view (holistic) is necessary to extract the information of the behaviour of complex systems. Trenhaile [9, 21] created new modules to explicitly incorporate waves and tides erosive mechanism like shear, abrasion or direct impact erosion. He also includes the beach morphodynamic on the erosive processes. Recently, the authors of this chapter [11] created a new PRM based on the idea of Kamphuis [26], and later Castedo et al. [27] developed another version of the marine treatment based on the work of Trenhaile [9, 21]. Castedo's models can solve cliff recession problems, including geotechnical stability analysis, in real profiles with variable lithology.

Usually, quality of the model results is established against field measurements and historic catalogues. Since the recession processes are regularly slow on a human scale, the designated study areas, where the results will be validated, must have a sufficiently extensive data record to provide the necessary contrast and allow model error estimation. As long as the catalogue has more registered values, the quality of the estimation error is improved. Also as the recession rate is higher, the sample rate (or field campaign to get the data) should also be higher.

In some cases, for certain lithology, it might think that the speed of recession is a criterion for discriminating those coasts that are not cliffs. However, this is not so. Although clay dominated coastlines change more rapidly than most rocky coasts, they share the fundamental characteristics that distinguish them from beaches; those are that erosion and coastal retreat are irreversible. In this case, we do not suppose the traditional distinction that has been made between clay and rocky coasts because it is untenable [21].

2. Coastal recession processes

In coastal geomorphology, the cliffs (**Figure 1a**) are defined as a geographical feature in the form of denuded coastal rock or cohesive soil moulded by the action of two processes acting at the same time. The marine processes act as a transport and erosion force below the water level. The sub-aerial processes determine the resistance force that produce the mass wasting phenomena [4, 28, 29], that is, landslides, topples, rockfalls, slumps, among others, of varying size distributions [20, 30]. Cliffs can be classified as follows: (a) steep to vertical slopes that

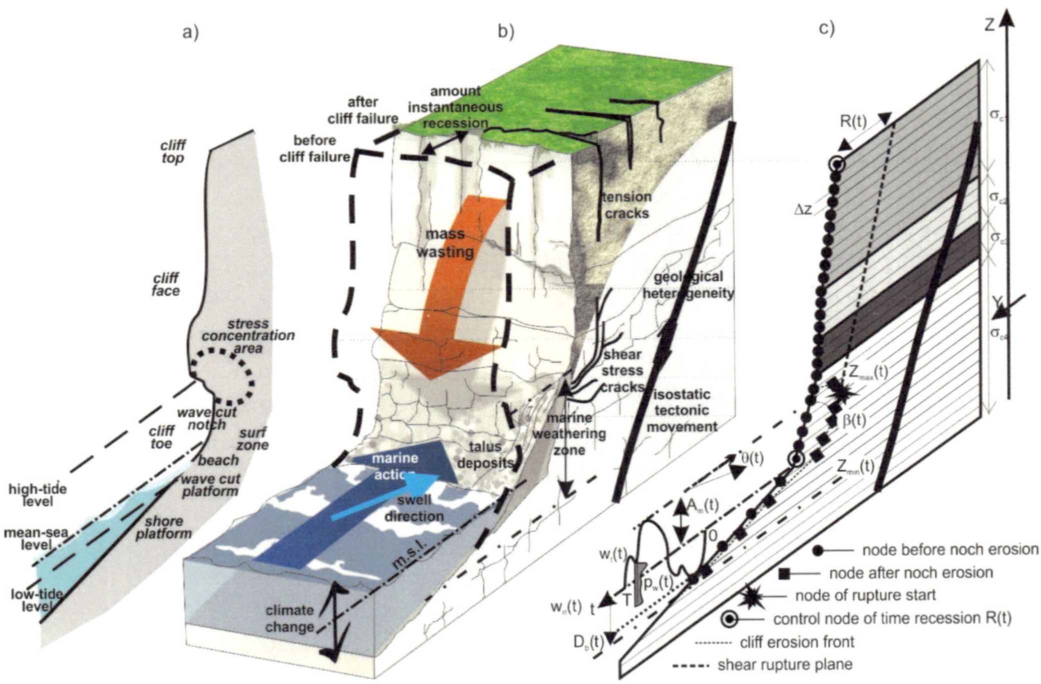

Figure 1. (a) Cliff morphological features; (b) cliff *behavioura*l unit (CBU) features and related processes and (c) spatial discretization and functions for process response numerical model.

expose hard or soft rock outcrops to the sub-aerial and marine weathering actions; (b) the bluffs that are cliffs of smoothed and rounded profile in which the exposed materials have a lower resistance and cover with vegetation and soil the deep rocky basement. According to Sunamura [4], three major morphologies have been distinguished on rocky coasts: the type-A shore platform or sloping type, the type-B shore platform or horizontal or sub-horizontal type, and plunging cliffs. The loss and erosion of the toe cliff by the wave's action are responsible for the growth of the two types of shore platforms, though no appreciable recession occurs on plunging cliffs.

The evolution of coastal cliffs is based on a balance between resisting and assailing forces. The lithology of the cliff material, principally described by its mechanical strength, determines the resisting force. The material resistance can be described by its compressive, tensile, cohesive or shear strength, being the first the most useful one [4]. The material resistance depends also on its formation conditions (bedding planes, cracks, joints, etc), the structure of the outcrops and latter of the human activity. The deterioration of its mechanical properties due to sub-aerial weathering (physical, chemical and biological) or wave attacks can enhance the cliff erosion. Although the marine weathering processes, such that abrasion, strain, solution, bioerosion, freezing and thawing, occur on a small scale and are relatively slow, their results shape and modify the coastal topography. These actions are most important at the base of the cliff (**Figure 1b**), where the impact of the waves and the abrasive action of the water loaded

http://dx.doi.org/10.5772/intechopen.68445

with suspended solids and fragments of rock is frequent. In this way, a basal notch grows and deepens, the overhanging rock mass increases concentrating the shear stresses in the bottom of the notch. Once the strength of the rock is exceeded, cracks and fractures arise mobilizing, detaching and collapsing the upper rock mass. The repercussion of the slow erosion processes on a small scale is translated on a large scale and suddenly down-slope movements in different ways (falls, topplings, slides, slumps, flows). Therefore, changes on coastal cliff morphology are not easily predicted because coastline recession is the cumulative result of successive numerous interacting gravitational phenomena, which are mostly coupled to marine hydrodynamics, intermittent and sporadic.

Observed complex behaviour arises naturally from several feedback patterns that control coastal erosion and recession, in the same way as has been usually presented in simple dynamical systems from physics theory. The system comprises two main components: the shoreline sub-system and the cliff sub-system. The influences on a coastal cliff system can be organized into a generalized hierarchy:

- Environmental activation mechanism: These are the elements that mainly control the recession problem like eustatic, vegetation, tectonic activity, isostasy, anthropic action, geology and climate
- Primary responses: They are the reactions (measurable) of the activation mechanisms that can produce changes in the coast like rainfall, wind regime, tidal-wave conditions, sea-level changes, morphology and geomechanical characteristics.

Coastal landsliding is a result of a combination of cliff toe erosion and geotechnical processes within the cliff, primarily connected with pore pressure distributions within the slope that influences stability [13]. Variations in groundwater level, which change the mobilized strength in soft materials, are significant contributors to the development of instability. The importance of groundwater level on the stability of coastal cliffs may be a significant uncertainty for predictions in view of a rapidly changing climate. In addition to long-term changes in groundwater levels, potential changes in surges size and frequency are an important consideration in model development [9, 22, 31]. The more rapid removal of protective debris aprons and changing water pressure distributions contributes to more rapid instability and erosion. Therefore, any model must accommodate changing climatic conditions in order to provide adequately resilient assessments of hazard.

Cliff recession phenomena release debris deposits at the cliff toe, which, at least temporarily, provides a natural protection against the sea forces or further erosion. Refs. [32, 33] have highlighted the importance of this protective colluvium effect on the coasts of California (USA) and Holderness Coast (UK), respectively. These authors emphasized the importance of long shore currents, which may remove such material from the cliff base towards embayment or stagnation areas where could be deposited. The losses of this front unconsolidated detached material allow erosion to continue. Front detached material, like debris or densely fissured rocks to soils, has its critical limit strength reduced due to the fracturing caused by combination of the mass collapse and the weathering. Additionally, cracks and specific surface increase because size distribution of rock and earth fragments is moved to smaller sizes, so it increases their erodability. Therefore, less front material is consolidated and size reduced, less is the time to be eroded and spent at the cliff toe.

The backshore, the foreshore and the nearshore are all affected by the processes of coastal cliff recession; this can be grouped into a single element called "Cliff Behaviour Unit" (CBU). Each CBU unit consists of a 3D block of cliff coast that can be conceptually simplified and represented as a vertical profile (**Figure 1c**) with uniform homogeneous geologic and oceanographic conditions. This concept provides an important framework for cliff management in order to tackle the study and modelization of the complex recession process. Each vertical section is a reflection of the interrelationships between the morphodynamic processes and resultant changes in form over time in the coastline, the backshore, the foreshore and the nearshore. Subsequently, the recession process in the shoreline can be computationally solved into different CBU units. Then, for each CBU is reasonable to calculate the balance of forces and sediments. Later, the relation of both balances can be used to make a quantitative analysis of the complex erosional-sedimentary dynamic regimes between CBUs. These units (CBUs) can be coupled to adjacent CBU's within the framework to form littoral cells, which will allow the study of several areas together with more accurate results. The relations between CBU's are important regarding the mass and energy flow exchange effects, that is, coastal landslides affecting the susceptibility of adjacent cliffs to further recession, sediments transport and

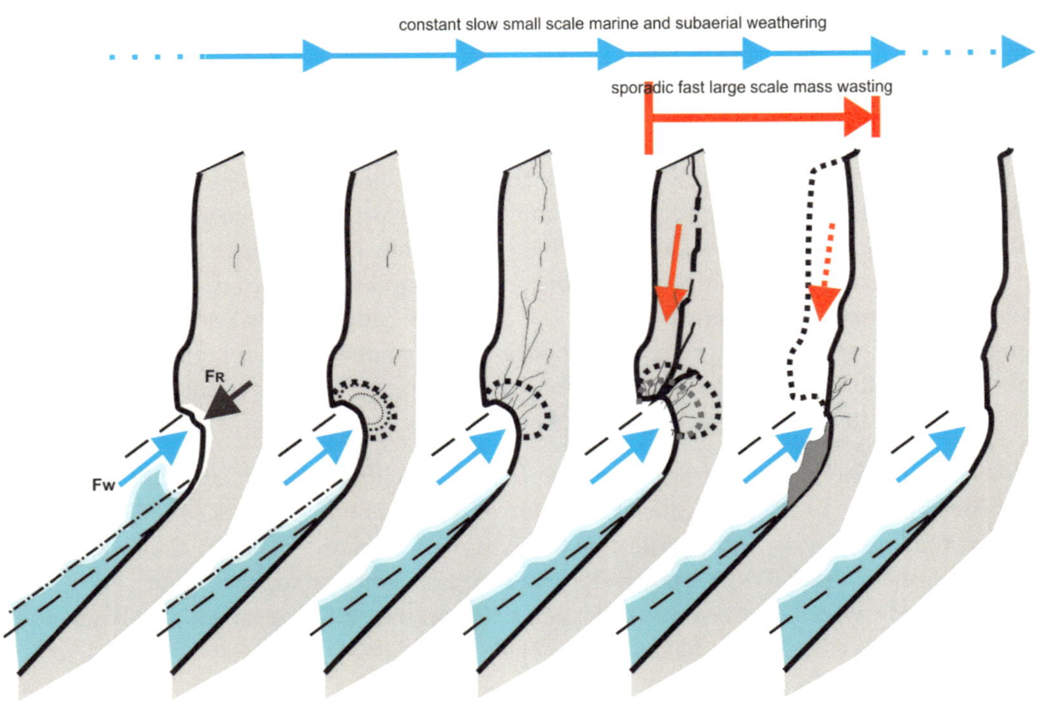

Figure 2. Schedule of the cyclic evolution of a cliff profile due to the persistent incidence of the assailing forces of waves Fw, the erosion in the wave cut notch cause Fw is larger than the resisting rock force of cliff forming rocks FR, and the sudden instability and subsequent mass wasting which produces the cliff face recession.

others. The activation mechanism and their primary responses that determine recession in a CBU can be schematically represented through cause-effect connections diagram (**Figure 2**) as a draft idea of the conceptual model acting on the whole CBU. This flow chart of interactions comprises soft and hard rocky coasts.

In summary, coastline recession on unprotected cliffs or bluffs is characterized by a cyclic process of marine weathering and abrasion of rock or soil particles through multiple physical chemical mechanisms. As consequence, the removal of material from the cliff toe, and the increase of depth at the wave cut notch. The deeper the notch, the more important the stress concentration around it [32, 34]. Then, if the material strength is exceeded, slope becomes unstable, and a mass failure occurs that retreats the coastline at the cliff top. This can reduce the coastal slope and deliver debris to the front beach or shore platform at the cliff toe. The resultant deposit may only be partly removed by marine action allowing an accumulation of colluvium to form which provides a passive support to the slope, and some, albeit small erosion protection. As cliff recedes, erosion penetrates though zones of weakness in the cliff face and top such as cracks or joints, cutting clefts and crevices that may develop into caves, blowholes, which probably will evolve into bridges, arches, promontories, pillars, as common natural rocky coastal geomorphologies. The complete removal of the debris deposit or protective structures at the foot or front of the slope results in the cycle beginning again.

3. Previous coastal erosion-recession models

In countless branches of science, computer models have helped in improving our understanding of dynamical behaviour in the earth physical processes. In the near shore, the design of computer algorithms tries to include process-related information to understand flow dynamics, particulate and chemical distributions, and the overall oceanography of coastal environments. Whether cliff morphodynamics is considered, most of the main characteristics and activation mechanisms are illustrated in **Figure 3**. Most of them have been described individually or jointly in the literature, that is, coastal cliff recession system, groundwater load, geotechnical characteristics, waves and tides characteristics [21, 35–37]. Some of the very complex characteristics and their primary responses have been solved mathematical and numerically [21–22, 26, 38]; individually or coupled, side by side (**Figure 2**). However, as presented before, a coastal recession phenomenon is a consequence of the interaction between the ground and sea forces at the point where they meet. Scientific results that adequately reproduce such a complex land-sea interaction have been scant and simplistic. Modelling results often presuppose that the recession pattern is the result of a single, unique distribution of waves, weather and environmental conditions. However, as different conditions, due to the range of behaviours found in different cliff morphologies (different CBUs), can cause diverse recession scenarios, the application of a simple prediction method cannot be universal [21]. Therefore, to develop a process-response model, it is crucial to take into account a wide range of information and existing feedback relationships and multi-scale coupling between the activation mechanisms and their primary responses that control the morphodynamics of the coastal recession events.

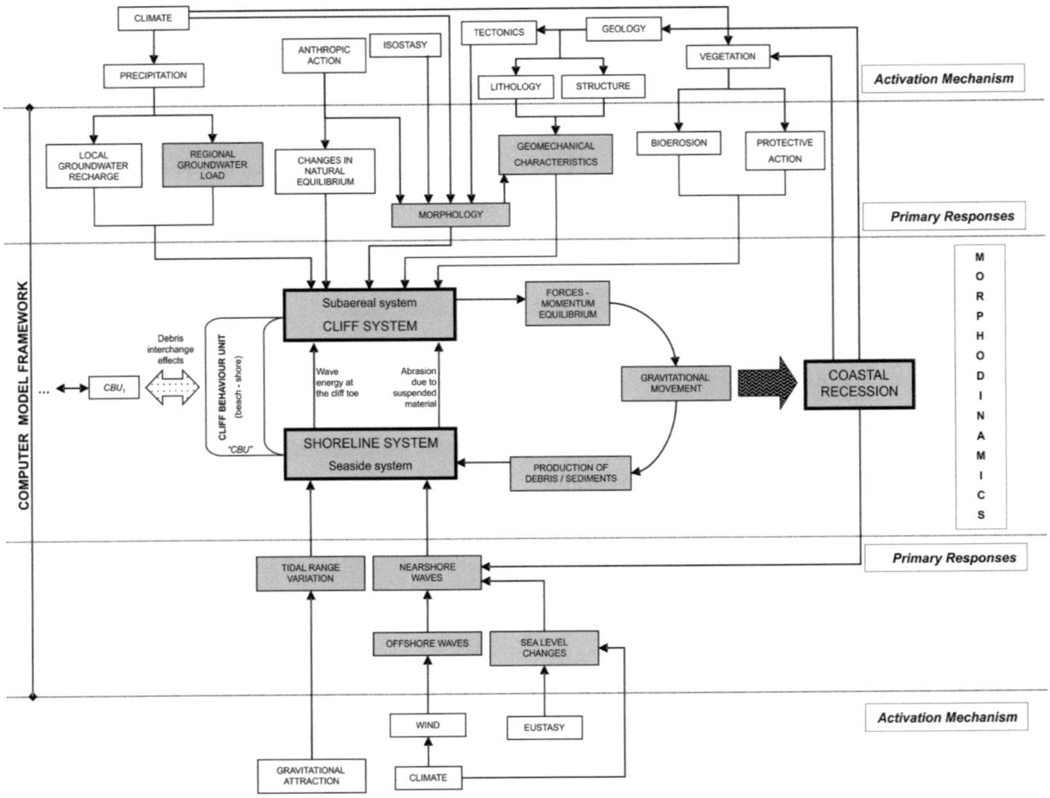

Figure 3. Flowchart with details of the activation mechanisms and their primary responses.

3.1. The earliest models

First attempts at modelling this complex interaction of processes were made by Sunamura [4] and Kamphuis [26]. Both authors considered the erosion at a point on the cliff as an average rate that quantifies the amount of material loss by coastal recession over time (usually one year). Yielded values, as the rate ($\delta(x,t)$) between marine driving and material resisting forces, could be considered as a safety factor from the geomechanical point of view [36, 37, 39] of the coastal erosion processes (Figure 5.2—chapter 5—[4]). Through laboratory experiments, some authors [40–42] validated closely first numerical approaches and realize that cliff change depends on spatial processes and time at different scales. Based on these data, and isolating the large number of elements involved in the recession event, an earliest model was developed. In this model, the change in position per unit time at a point x in the cliff face (erosion) is a function of: (i) time (t)—that control the number of tidal cycles that act a certain event; (ii) the assailing forces (F_W)—hydrodynamic and hydrostatic; and (iii) the resisting forces (F_R)—that depends on the lithology and structural conditions. The model is then based on the force ratio during a certain time (t) and quantifies the displacement of each point of the coastal section as, Eq. (1):

$$\delta(x,t) = f(F_W(x), F_R(x), t) \leftrightarrow \frac{dx}{dt} \approx \frac{F_W(x,t)}{F_R(x,t)} \quad (1)$$

Various attempts have been made to associate F_W and F_R with measurable, sea and rock, respectively, characteristics. Among others, the work of Craig [26] provided an expression in which the material strength is represented by a factor to be calibrated experimentally. Subsequently, the model by Sunamura [4], derived from experiments on a concrete wall in an artificial wave tank, suggested that F_R is directly proportional to the uniaxial compressive strength (UCS) of exposed rock. Also Sunamura [4] suggests that the erosion rate at a certain time $\delta(x,t)$ is proportional to the logarithm of the ratio between both forces, F_W/F_R, applied at the position x. Resistance forces in this context means the strength and hardness of rocks attacked by the physical forces of marine erosion, or their durability in the face of physical, chemical and biological weathering process. Rock resistance depends on several factors such that lithology, joint or fracture density, and the existence of bedding or cleavage planes which decrease rock hardness or soil cohesion. Identifying the most appropriate parameter to represent the strength of the material to erosion has been investigated in numerous laboratory studies suggest the compressive strength σ_c (Japan island: andesite lava, sandstone, clays, gravels, among others—[4]; Cilento area: sand, calcilutites and siltstones—[38]) the most appropriate parameter. Although less frequently, the tensile strength (for idealized cliffs in chalk—[34]), cohesion (Calvert Cliffs, Maryland: alternation of diatomaceous earth, sand, clay and marl—[43]), shear strength (Lake Erie: glacial silt/clay overlain by lake sand—[26]; Grates Lakes: cohesive clay coast—[21]) or Young's modulus (Fukushima coast: gravel, sandstone, mudstone—[44]) have also been proposed. Nevertheless, the model that most completely describes the resisting forces is the Budetta et al. [38]. The authors introduce into Sunamura's [4] model the Rock Mass Index (RMi), taking into account not only the UCS value of the intact rock but also taking into account the block volume and the fracturing state (joint condition factor). Their results show this parameter capture the relationship between the resisting forces and their geotechnical controls. Incorporating a suitable representative form of rock mass (i.e., the rock material plus the fractures) was an important step forward in the prediction of recession.

3.2. Historical recession based models

Historical trend analysis [24, 45] uses the expression based on the 'Bruun Rule' [46] and modifications to the Bruun rule [13], on two-dimensional profile models, to predict future shoreline erosion depending on projected future sea-level rise scenarios and measured shoreline recession rates. The main idea behind the Bruun Rule is the statement that the upper shore face (the morphologically active—years—part of the continental shelf) has a profile that is in equilibrium with the hydrodynamic forcing (i.e., wave driven processes mainly, but also density driven currents close to river or estuarine entrances) and that for any perturbation in the forcing, the reaction of the profile will be moderately fast. When sea level rise occurs the shoreline retreats and a new equilibrium profile will form at the new shoreline position by moving sediments to deeper water, or in other words, the profile is shifted to an upward and landward position. Bruun [46] developed a model that evolved as a result of sea level rise (S) until it reached the shoreline retreat equilibrium, $R = S{\times}L/(h{+}B)$. Where L is the cross-shore width of the active profile, h is the closure depth, and B is the elevation of the beach or dune crest. Even though the Bruun Rule is very often used, it has a number of limitations, and care must be taken when using this tool to predict the coastline response to sea level rise. It basically tends to overestimate the future recession rate.

The following equation [24, 45], based on the Bruun Rule, predicts future shoreline erosion by using projected future SLR scenarios and measured shoreline erosion rates $R_2 = S_2{\times}(R_1/S_1)$, where R_2 is future shoreline recession rate (m/year); S_2 is projected future sea-level rise rate (m/year); R_1 is measured shoreline erosion rate (m/year), and S_1 is sea-level rise rate (m/year) during the period of measured shoreline erosion. The R_1 could be obtained from the Digital Shoreline Analysis System (DSAS) [47] fitting a least-squares regression line to all shoreline points for a particular transect. The historic rate-of-change of position of cliff-top is the slope of the linear regression (m/year), also known as historic rate of change. Recent approaches to historical shoreline analysis have been made using the Digital Shoreline Analysis System (DSAS) version 4.3 [47]. Additionally, statistical models [14, 24] should be used to predict future shoreline locations using historical information extracted from the DSAS. Within DSAS, the linear regression method is usually applied to obtain rates of coastline change statistics, computationally done in a GIS environment, for a time series of shoreline vector data that describe temporal evolution of the coast from multiple shoreline positions [47]. This method is selected because it includes the following features: first, all the data are used; secondly, it is also the most commonly applied statistical technique for expressing rates of change [48]; and thirdly, it includes additional statistical information essential to be compared with the yield results from other analytical methods (i.e. [14, 24]).

3.3. Probabilistic forecast methods

A key feature of these methods is the recognition of the episodic nature of cliff recession at many coastal sites. In other words, cliff recession proceeds primarily via occasional landslide episodes, non-uniformly distributed in size, followed by periods of relative inactivity, which may last for more than 100 years on some coastlines [49]. The process is complex and far from purely random, so maybe probabilistically distributed in some cases.

This methods use to present significant limitations, but on the contrary, the projection of historical rates is probably the most evident and straight forward approach. However, the historical record uses to consist of a less than 10 measurements over the last 100 years. This is insufficient to explain the pattern of recession events, and so the cumulative process of erosion, rock tension, and finally failure. In this case, the historical record can only reveal a limited picture of the past events.

The scarcity of historical cliff position data can limit the usefulness of many conventional statistical methods, such as linear regression. One approach to addressing this problem is the development of probabilistic models [50] to simulate the recession process. These models consider the cliff recession as an episodic random process and use a Monte Carlo sampling tool. This model can be used to simulate synthetic time series of recession data, which conform statistically to the cliff recession measurements. If the cliff recession process is plotted against time, its episodic form is reflected in a stepped function. If the simulation is repeated several times, a probability distribution of cliff recession can be built. The simplest model can be defined by two distributions: a description of the timing of recession events, and an event size distribution describes the magnitude of recession events in terms of the mean size and their variability.

For different cliff settings, a different probabilistic method can be used. The selection of the method should be based on: the knowledge on cliff recession of the study area, the nature of available data, if the primary responses are expected to remain constants and the amount of investigation and analysis which can be justified.

The type of information available is influenced by the nature of the cliff recession process. Also the results of any model will be influenced by the information available. For these reasons, more than one method must be used to indicate, in some way, the robustness of the predictions. Even in some cases, as the quality and quantity of data increases, more sophisticated analysis must be re-done.

3.4. Process-response models

The factors and phenomena contributory to development of many different geological processes can be systematized in term of process-response (PR) or cause-effect models. In the case of sea cliff systems, they are based on experimental-theoretical knowledge and mathematical description of the interactions between the hydrodynamic characteristics, the coastal processes, geotechnical properties and mechanical behaviour of the rock mass. Such models are still at an early stage of development, and most were originally purely deterministic. However, in order to incorporate uncertainty in the characteristics and variability of physical phenomena, a parametric probabilistic framework can be included addressing the complexity of the system using simple stochastic models of each considered process. In these models, as in previous ones, historical records were used for calibration. Also, in this type of model, most of the physical interactions are studied at laboratory scale so upscaling them remains problematic.

Walkden and Hall [22] developed a numerical model, SCAPE, to simulate the processes on a cohesive shore profiles. SCAPE is mainly based on the model of [26]. However, they added the effects of waves by a cumulative effect of an erosion function per wave developed through a tidal period, and the effects of coastal development platform, including changes in shore slope. The mathematical solution adopted means that the shore slope is limited to 10°in order to avoid numerical instabilities that produce non-natural roughness. Their erosive function was based on erosion experiments by Refs. [41, 42]. The weakness in this model is that it does not consider the characteristics of the rock, so that the resistive capacity of the ground is adjusted through the calibration constant based on the availability of historic recession data. Similarly, the failure criterion used is a deterministic approach, after 10 recession events, the model removes any overhanging material from the most eroded point in the notch, which produces a vertical cliff. This severely limits the use of the model, since the time when a break occurs or not, greatly affects the final appearance of the erosion profiles.

Trenhaile [21] presented a model to examine the relationships between variables that affect the erosion and development of cohesive clay coasts. Trenhaile introduces a relationship to consider the effect of the abrasion due to the suspended material. The erosion rate within the model is introduced as the amount by which the applied bottom shear stress produced by the waves, exceeds the critical shear stress for erosion. This model produces shore-normal submarine profiles retreat but it not reproduces the upper part of the section that is the cliff face.

4. Coastal cliff model development

The process-response models (PRM) are the most modern and updated software to incorporate critical activation mechanisms and subsequently trigger the primary responses in a CBU. In order to reduce the design complexity in the model, the process in the cliff-shoreline system is represented by single modules as in **Figure 3**. Each one is described mathematically by a single numerical approach to improve computational efficiency, high accuracy of results and assist understanding of emergent properties of the systems behaviour.

The recession model described here is validated in coasts with rocks with a compressive strength σ_c that varies from very soft (geotechnically named very weak, i.e. <1.25 MPa) to moderately hard (i.e., weak to moderately strong, 1.25–50 MPa). These kinds of rocks are also known as soft erodible materials. After the eroding process explained previously, the crest of the cliff moves inland a quantity $R(t)$, which usually varies at each failure event. This is the point of interest in determining the risk to cliff-top assets. So, at this point here is located control, and monitoring observations are regularly collected to determine the instantaneous position of the cliff, which is the point. For more details of model applicability and development see Refs. [11, 12].

The model explained here is the first one that explicitly includes a slope failure mechanism using a limiting equilibrium procedure in a dynamic manner, Fellenius' method of slices [39]. Previous models like [21, 22] uses an expert judgement criteria where the authors assume that each failure event occurs at every ten recession events, or a probability distribution to describe the possible cliff top location following a failure event [51].

4.1. Erosive model details

As outlined previously [11, 12] in this numerical model, the rock profile, $y\ (z,\ t + T)$ represented as a column of horizontally aligned layers of height $\Delta z = 0.05$ m, moves a quantity $\delta y(z,T)$ in the points z affected by the erosive processes after one tidal period T. This quantity can be defined as the erosion of each element of the cliff front and depends on $H_{b,}$ which is the breaking wave height; $T_{b,}$ which is the wave period; K, which is a calibration term representing hydrodynamic constants (100 $m^{13/4}s^{7/2}/kg$ for the application site; see Refs. [11, 12]); $\sigma_c(z)$, which is the uniaxial compressive strength (USC) of the rock mass; $\partial_z y(z,t)^{-1}$, which is the slope of each rock element z (which changes through simulation time in response to the calculated erosion and consequently to profiles evolution); $p_w(z,t)$, which is the shape function (which include from experimental data—[41, 42]—the erosion of different type waves) and $w_t(t)$, which is the tidal expression introduced as a sinusoid that oscillates about mean sea-level with an amplitude A_m and period T (both can be obtained through governmental marine agencies). The rock profile evolution can be described as seen in Eq. (2):

$$y(z, t + T) = y(z, t) \pm \frac{H_b^{13/4}(t)\, T_b^{3/2}(t)}{K\, \sigma_c(z)} \left(\frac{\partial y(z, T)}{\partial z}\right)^{-1} \int_0^T p_w(w_t(t) - z)\, dt \qquad (2)$$

The *in situ* cohesive material can be represented by its lithological characteristics if a field survey has been carried out. They are incorporated into the model using the value of the uniaxial compressive strength $\sigma_{\text{c-mass}}(z)$ of rock or soil mass calculated at each elevation z. In order to estimate the value of UCS, $\sigma_{\text{c-mass}}(z)$ of rock mass, the Hoek-Brown failure criterion was used based on rock mass rating (RMR) and estimated values of the constants m and s [52]. Alternatively if the coefficient of internal friction angle ϕ', and the cohesive strength c' values for the Mohr-Coulomb constitutive model are known it is possible to estimate the value of UCS $\sigma_{\text{c-mass}}(z)$ of the rock mass (Eq. (2) from [52]). In addition, the original material of the cliff composes the colluvium, but their original soil structure can be modified controlling the mechanical properties of this new structure. So the considered value of UCS for the colluvium $\sigma_{\text{c-weathered}}(z)$ can be obtained through the UCS for remoulded materials σ_{cR}, the internal friction angle ϕ'_w, and the cohesive strength c'_w, following the same methodology as before. In both cases, when the necessary data to obtain the values of UCS $\sigma_{\text{c-mass}}(z)$ or $\sigma_{\text{c-weathered}}(z)$ following the method presented by [52] are unknown, the considered values are σ_c and σ_{cR}, respectively.

4.2. Extra modules (primary responses)

Erosion, cliff stability and sediment transport are calculated one per tidal cycle T which is considered as one time step in the computational procedure of the global simulation time. As the foot erosion progresses, the notch becomes deeper and deeper increasing the volume of the overhanging material. Each tidal cycle geotechnical equilibrium conditions are evaluated to determine the stability of the exposed rock mass. This stability will be higher or lower as a function of the fracture grade of the rock mass, the material strength and the groundwater level. The first two parameters are introduced into the model by the RMR. When the retreating forces overcome the resisting ones, the rock strength is exceeded, resulting in the material critical failure (that can be deposited or removed) causing the loss of some coastline and consequently coastal recession.

Usually, for practical purposes, the groundwater conditions are based on the position of the groundwater table, ranging from fully drained to saturated. It is defined by the ratio where the phreatic surface coincides with the ground surface at a distance behind the toe of the slope among the slope height [37]. In this model, the water table elevation is described as the Dupuit parabola based on the upstream head, measured from the impermeable base, and the downstream head data, see **Table 1** for details [35]. The inland pressure is considered here as a difference of the depth from the surface to the seepage point b, minus the depth from the surface to the groundwater level a. The downstream pressure is considered as the reference zero-level at the spring point, see **Figure 4**.

Clay cliffs are susceptible to deep-seated rotational landslides, which occur where basal erosion is rapid enough to remove the debris and steepen the underlying coastal slopes [28]. This model utilizes limit equilibrium techniques; for simplicity, accuracy and computational time, the ordinary method of slices (OMS) technique is used [39]. The OMS method was developed

Factor of safety to slumps (FSslump)	$FS_{slump} = \frac{\sum_{i=1}^{n}(c'_i \Delta l_i + (W_i \cos\theta_i - u_i \Delta l_i)\tan\phi'_i)}{\sum_{i=1}^{n} W_i \sin\theta_i}$
Variation in groundwater elevation (h)	$h(y_D) = +\sqrt{(y_D/L)(b-a)^2}$
Maximum bending stress (σt max)	$\sigma_{tmax} = M/Z$
Bending moment (M)	$M = 1/2\gamma d\, h_{failure} L_n^2$
Modulus of the section (Z)	$Z = 1/6 d\, h^2_{failure}$

where n is the number of slides, i is the i slide, c_i' and ϕ_i' are the effective cohesion and friction angle, respectively, Δl_i is the length of the slide, W_i is the weight, θ_i is the inclination angle of the bottom of the i slide, u_i is the pore water pressure at the centre of the base of the i slide, y_D is the horizontal distance to the seepage point to any point in the parabola, L the horizontal distance between the seepage point and the point where the phreatic surface was measured, b the seepage point, a the phreatic level measurement,γ is the material unit weight, d is the distance along the bluff face of the overhanging material, $h_{failure}$ is the height, and L_n is the notch depth.

Table 1. Equations to calculate the stability at every calculation time step T [11, 12].

by Fellenius in 1936 and is sometimes referred to as "Fellenius' Method." Details of the equation used and the schematic representation can be seen in **Figure 4** and **Table 1**. In the OMS method implemented in this model, the failure circle is defined by the position of a point through which the circle must pass at the toe of the slope and the tangent angle (α) to the circle plane at this point. Also is necessary to define the point at which the circle intersects the cliff top (point 2 in **Figure 4c**). The variation of the second point (2 to 2′) in the x-axis allows the code to look for unstable circles. The maximum depth, from the cliff edge, to look for an unstable circle is h_{talus} based on physical observations [36, 39]. Also, different circles are created by separating 2 to 2′ a discrete quantity Δc selected by the user. If the value of factor of safety (FS) for each slip surface is less than 1.0, slope is unstable. The factor of safety obtained with the "Fellenius' Method" is the lowest when comparing with other techniques, so the results obtained with this method are on the safe side [39].

In addition to the OMS method, this recession model includes a notch collapse stability analysis based on the ideas proposed by Refs. [53, 54]. The cantilever beam model derives the distribution of stress inside the cliff assuming that tensile stress acts on the upper section and compressive stress on the lower section, see **Figure 4d**. The maximum bending stress ($\sigma_{\text{t-max}}$) is calculated as shown in **Table 1**. The tensile strength (σ_{t}) is specified and limited, which in many analyses is taken to be 10% of the rock mass cohesion [37], for isotropic materials. Then, a collapse occurs when the maximum bending stress ($\sigma_{\text{t-max}}$) exceeds the tensile strength of the rock (σ_{t}).

Following a failure event disintegrated collapse material is deposited at the foot of the cliff, which acts as a natural protecting material. This talus material now located at the foot of the slope is more altered and fractured than the original one, and, therefore its resistance to erosion is reduced, this effect becomes more noticeable at smaller values of $\sigma_{\text{c-weathered}}(z)$. It is usually quickly removed by the cumulative action of waves and currents. In the model, debris

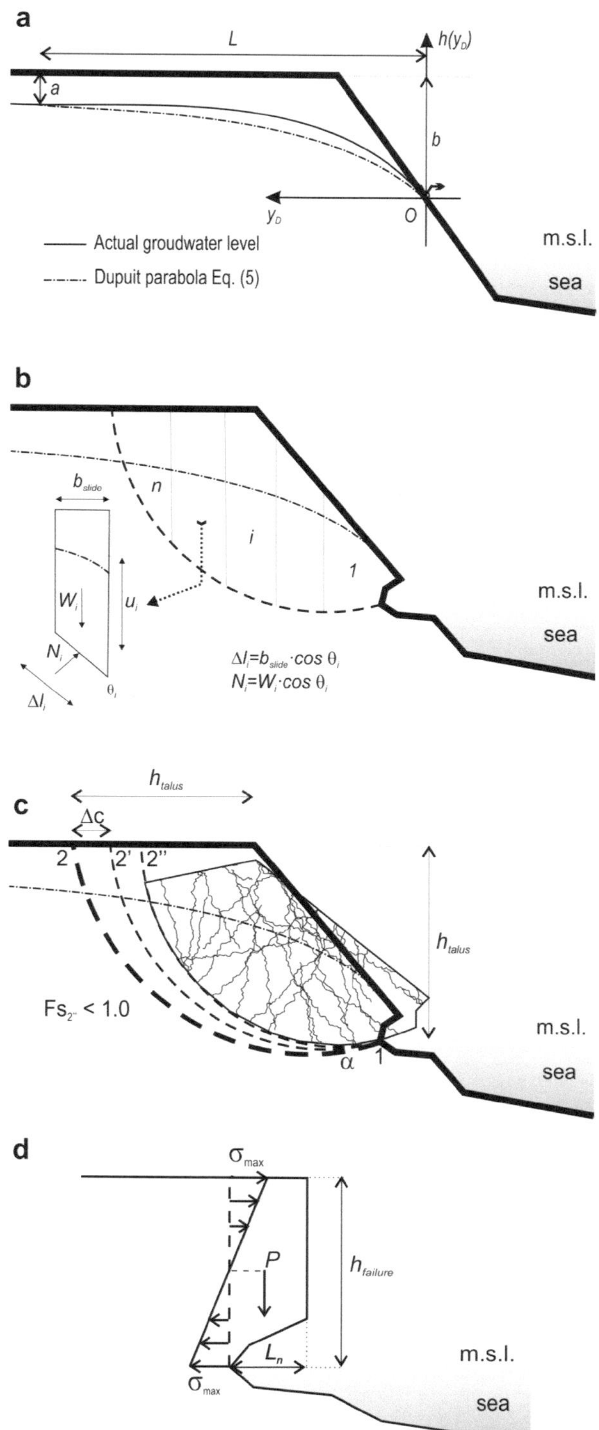

Figure 4. (a) Schematic cliff section showing the adapted Dupuit equation solution for flow through a porous media; (b) schematic visualization of the factor of safety calculation following the Fellenius method; (c) search criteria for circular arc slip surfaces from (b); (d) schematic visualization of the maximum bending stress situation.

produced from the cliff face are deposited on the colluvium slope and covers the foot of the cliff, in the proportion defined by the user. Three different solutions are developed: the colluvium shape is determined by the friction angle for weathered materials ϕ'_w [36, 37, 39]; the angle of reach for landslides, defined from the detaching point with respect to the horizontal, as the ratio of vertical height differences between the source point and the furthest point reached in distance to the horizontal [55, 56]; and based on expert field observations where the talus is created from the half cliff height ($h_{talus}/2$) with a small slope [7, 28].

5. Model application: Holderness Coast (UK)

This model has been applied within the cliffs of the Holderness Coast (UK), which stretches from Flamborough Head in the north to Spurn Head in the south, presenting a smooth "S-shape." The area is sparsely populated, and land use is predominately agricultural and tourism; however, critical infrastructures (transport and communication networks) are located near the coast [33] (**Figure 5**).

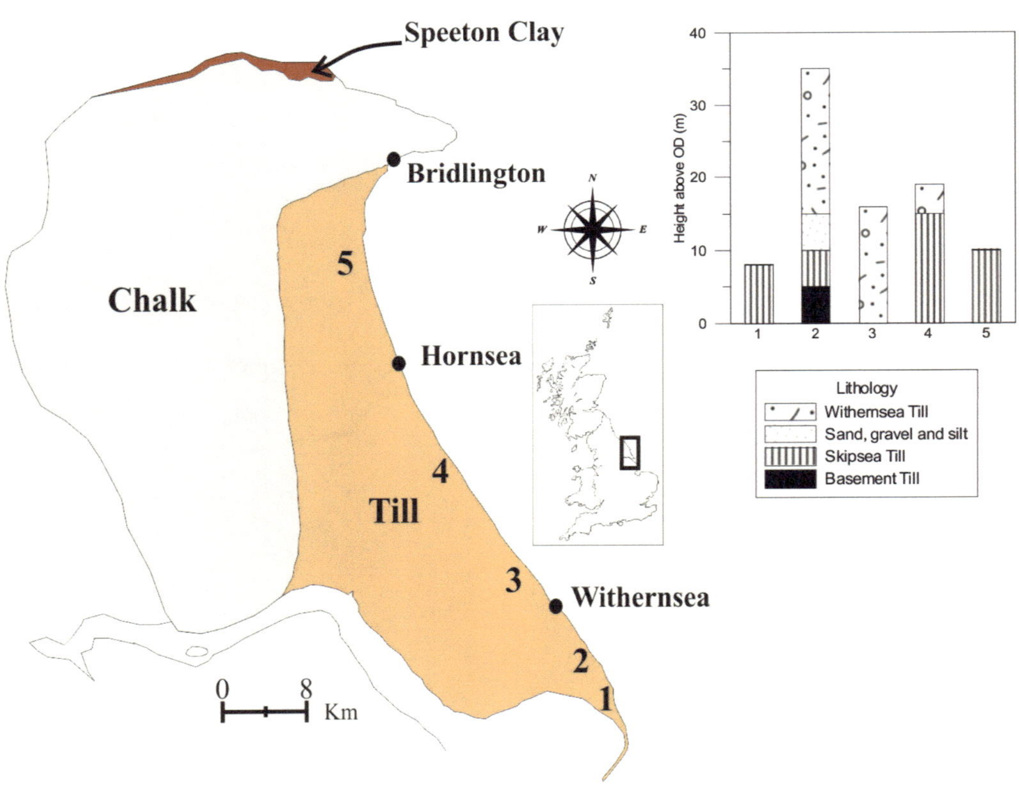

Figure 5. Location and geological sketch of the Holderness Coast. Note that, the numbers are to situate the lithological columns through the area.

This area is one of the youngest natural coastlines of England, a 61 km long stretch of low glacial drift cliffs ranging from 3 to 35 m in height. The Holderness Coastline is made up of soft boulder clays (tills) left after the retreat of the Devensian ice sheets about 12,000 years ago. These soft, recent deposits sit on a platform of chalk which slopes away gently to the east originating from between the Santonian (85.5 Ma) and the Maastrichtian (65 Ma) stages from the Upper Cretaceous period. The chalk is located at a depth of at least 30 m below the commonly used UK datum (Ordenance Datum Newlyn—OD), across the vast majority of the coastline [57]. The most recently and accepted division of the glacial materials was summarized by [58], who divided the area into: Basement, Skipsea, and WithernseaTills. The oldest unit (Basement Till) is the lowest of the stratigraphic series. The following of the series is the SkipseaTill, and finally, the Withernsea Till is found. In addition, the majority of the coastline is also capped by a weathered till deposit. The cliff stability alongside the Holderness Coast has been the subject of study and controversy over the last 30 years. Field and laboratory testing have shown that the tills display essentially similar geotechnical characteristics used as an input parameter (see Refs. [11, 59–61]).

At Holderness, the environment is mainly dominated by wind wave development. The dominant wind direction is north-easterly, determining also the wave direction, and creating a north-south orientated net long shore current. During storm events, highest waves can reach up to 4 m. In addition to this, tidal range at the Holderness coast is very high and can reach up to 7 m.

Since the Roman times, many villages and parishes have been lost [8], with an estimated area of more than 200 km^2. Along the entire coast [33] an average recession rate of 1.55 m/year has been recorded, making this coast one of the fastest eroding coastlines of Europe. Due to this erosion rates, the East Riding of Yorkshire Council designates important economic resources to collect cliff top retreat and coastal cliff profiles measurements within the entire coastline since 1951. A common conclusion for the rapid erosional behaviour of the Holderness Coast is the low cohesive values of the glacial till and the groundwater.

The model requires input values for oceanographic data, sea level rise estimations, retained material that form a talus after a failure event, the geotechnical data and the groundwater level. Wave height in deep water (H_0 = 0.98 m) and mean breaker period (T_b = 7.66 s) are entered as a mean values from 3 years data registered by the Hornsea Directional Waverider Buoy, deployed on the 05/06/2008. The tidal amplitude (A_m = 3 m) and tidal period (T = 12.46 h) are obtained from The National Oceanography Centre. The hydrodynamic constant (K) had a value of 10^2 for the entire Holderness coastal area. The groundwater level is also an important factor in the entire system with special influence on the failure mechanism. For modelling purposes of the Holderness area, the groundwater level used is 2.5 m under the surface at about 250 m landward. The seepage point can vary from one profile to another and usually require specific site investigation [33].

A changing climate and particularly an accelerated sea-level rise are believed to impact cliff retreat rates [3] with an expectation that shoreline retreat rate will generally accelerate in the future [11–13, 16, 17]. According to United Kingdom Climate Impacts Programme (UKCIP), relative net sea level will rise between 0.002 m/year and 0.01 m/year for Yorkshire by the 2080s for the low emissions and high emissions Scenarios, respectively [62]. However, Walkden and Rossington [63] recommend that for planning purposes, a relative rise of 0.006 m/year is taken into account.

Profile	LRR (m/year)	LCI (m/year)
P21	1.83	0.63
22N	1.73	0.45
P22	1.76	0.24
23N	1.73	0.31
P23	1.65	0.44
24N	1.57	0.70
P24	1.46	0.54
25N	1.43	0.45
P25	1.51	0.60
26N	1.41	0.56
P26	1.46	0.69
27N	1.37	0.84
P27	1.20	0.62
28N	1.10	0.51
P28	1.13	0.68
29N	1.08	0.77
P29	1.05	0.57
30N	1.13	0.41
P30	1.05	0.39
31N	1.17	0.34
P31	1.14	0.44
32N	1.18	0.43
P32	1.10	0.24
33N	0.95	0.26
P33	1.03	0.26
34N	1.10	0.23
P34	1.13	0.30
35N	1.13	0.25
P35	1.10	0.23
36N	1.02	0.32
P36	0.95	0.28
37N	1.07	0.44
P37	1.03	0.32

LRR, linear regression rate; LCI, standard error of LRR at 99.

Table 2. Historical recession rates for several profiles measured in the field (noted as P followed by a number that follows the numbering provided by Ref. [64]) and artificial profiles obtained from DSAS analysis (that follows the same numeration as the real profiles with the addition of a letter N) for the 159 year period 1852–2011.

5.1. Historical analysis based on DSAS

Table 2 shows the historical recession rates of change calculated using the real profiles provided by the East Riding of Yorkshire Council [64] and the artificial profiles created in ArcGIS to carry out this work for the 159 year period 1852–2011. The minimum R^2 value of these regressions is 0.89, see Ref. [12] for more details. The recession rates vary from one profile to another with values from 0.95 to 1.83 m/year, making the erosion trend so irregular. These variations can be due to numerous factors such as sub-horizontal or vertical discontinuities, cliff height disparities or shore platform irregularities.

5.2. Models output

Another way of checking the goodness of the PRM developed is comparing its output, with shorelines predicted by Lee and Clark [14] and Leatherman [24] models. Both models are used with the data extracted from the DSAS software. The PRM has been run with two different scenarios, one with a SLR of 0.002 mm/year to cover the lower prediction, and the other one with a SLR of 0.006 mm/year to cover the opposite (**Figure 6**). After that, future recession values are calculated in several epochs (2011, 2021, 2031, 2041 and 2051), and a shoreline was estimated in each one.

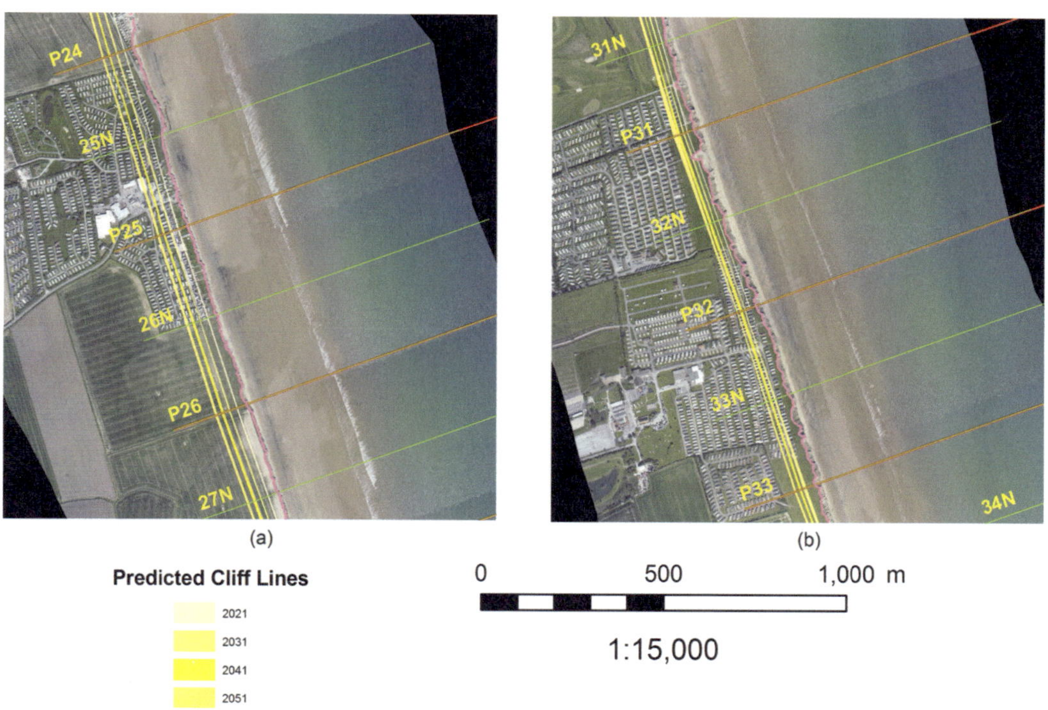

Figure 6. Predicted cliff lines for PRM: (a) Skipsea Sands; (b) Far Grange region. The cliff edge marked line represents the 2011 cliff-top position.

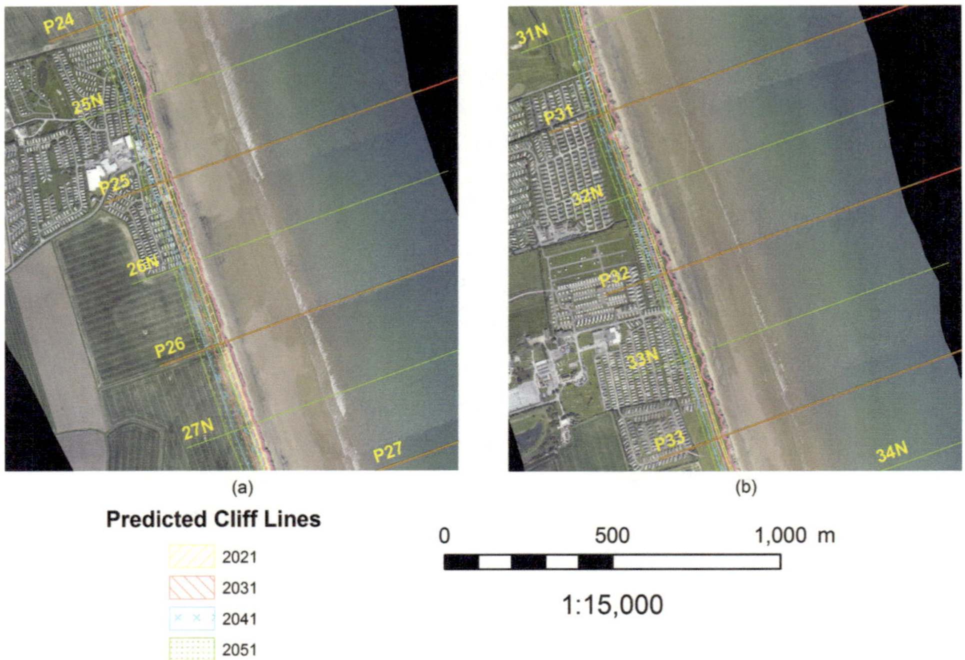

Figure 7. Predicted cliff lines for Leatherman model: (a) Skipsea Sands; (b) Far Grange region. The cliff edge marked line represents the 2011 cliff-top position.

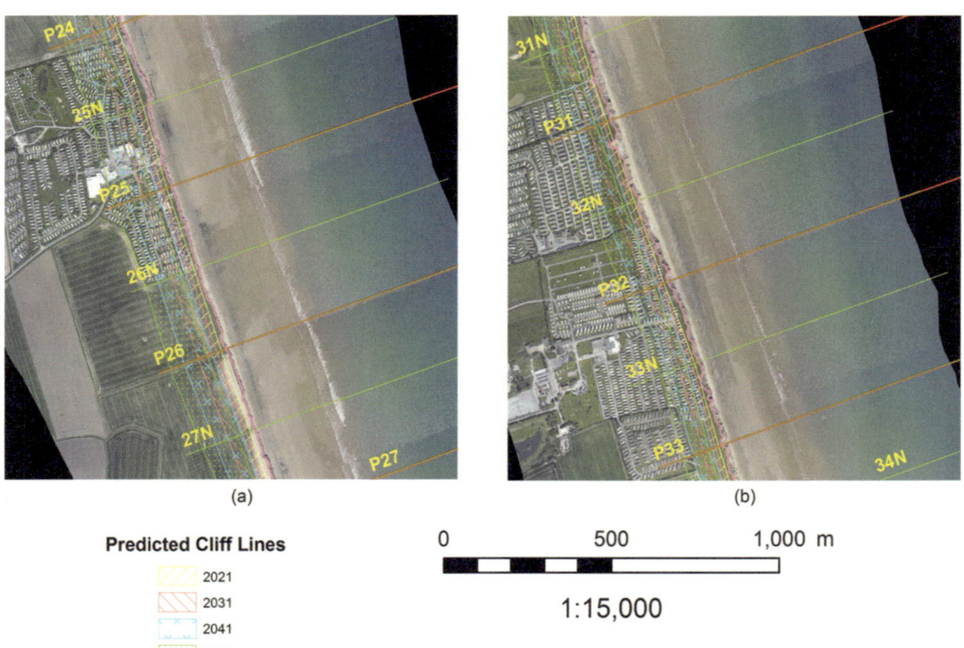

Figure 8. Predicted cliff lines for Lee-Clark model: (a) Skipsea Sands; (b) Far Grange region. The cliff edge marked line represents the 2011 cliff-top position.

The Leatherman model [24], also called historical trend analysis, uses the equation presented in the point 3.2: $R_2 = S_2 \times (R_1/S_1)$. Here, R_1 is the average recession rate (LRR) and S_1 the historical sea level rise equal to 0.002 m/year [1]. Finally, the S_2 is the projected future SLR considering two scenarios: 0.002 m/year (lower bound) and 0.006 m/year (upper bound), see **Figure 7** for details.

The Lee-Clark model [14, 17] calculates the future recession as follows: recession by year A = (LRR ±LCI) × T. where LRR and LCI values are obtained from the DSAS analysis and for the study case can be found in **Table 2**. To cover the upper and lower bounds, the LCI can be subtracted or added to the LRR value, respectively. T is the time to extend the simulation. The output can be seen in **Figure 8**.

5.3. Models comparison

The study presented here has quantified the recession of part of the Holderness Coast shoreline, from profile P21 to P37, located between Bridlington and Hornsea. The recession values obtained with the three models presented here (**Figure 9**) are in accordance with data obtained by recent analysis based on historical data for the same location [10, 65]. The Leatherman model presents the highest recession rates and higher uncertainties as the shadowed region (between lower and upper bound) is the biggest. The Lee-Clark model, based on historical analysis is the most optimistic in terms of low erosion; however, it is based on hundred years

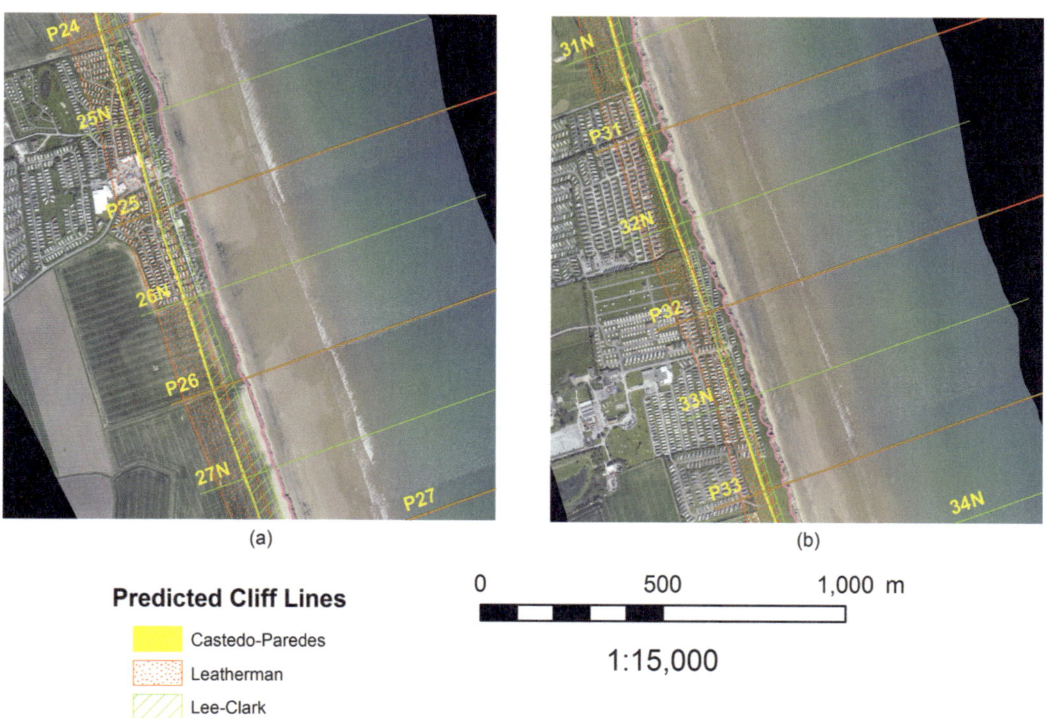

Figure 9. Predicted cliff lines for PRM, Leatherman and Lee-Clark models at year 2051: (a) Skipsea Sands; (b) Far Grange region. The cliff edge marked line represents the 2011 cliff-top position.

of data. In the middle, we found the PRM model which is less sensitive to the change of the sea level as higher values can decrease the cliff face height increasing its geotechnical stability. Also, this latter model respond better to other changes that may arise during time like changing in profile shape, marine conditions, etc. This empirical model can provide deterministic predictions of recession but in reality cannot reflect information of the process variability and uncertainty. Obviously, this is a problem as cliffs with the same recession rates can have a totally different behaviour. On the contrary, these models are of friendly use and "only" need historical data, while the PRM model needs geological, marine, topographical and also historical recession data.

6. Final statements

The estimation of a future shoreline is a complex problem in engineering, due to the stochastic nature of cliff retreat, which is apparent from the irregular cliff lines. The most widely models used to forecast erosion is through use and analysis of historical records. The methods to predict the future cliff-top evolution exposed here rely on the knowledge of historical recession rates, especially the empirical models, such as the Lee-Clark model [14] and the Leatherman model [14]. The model presented here has been applied so far on shores with soft, erodible material, where erosion is dominated by wave action; however, with the introduction of the rock mass UCS, the model range applicability can be extended to hardest materials. The review and results provided here can thus be used to help the developers in the process of deciding how to manage the potential losses (social, ecological and economic), and when and where to intervene in the erosion processes at least for rapidly eroding coasts like Holderness, UK.

Author details

Ricardo Castedo*, Carlos Paredes, Rogelio de la Vega-Panizo and Anastasio P. Santos

*Address all correspondence to: ricardo.castedo@upm.es

Department of Geological and Mining Engineering, School of Mines and Energy, Universidad Politécnica de Madrid, Spain

References

[1] Eurosion. Living with Coastal Erosion in Europe: Sediment and Space for Sustainability. Results from the Eurosion Study; European Commission, Luxembourg, 2004. 40 pp

[2] Nicholls RJ, Cazenave A. Sea-level rise and its impact on coastal zones. Science. 2010;**328**(5985):1517-1520

[3] IPCC. Climate change 2014: Mitigation of climate change. In: Edenhofer O, et al., editors. Contribution of Working Group III to the Fifth Assessment Report of the Intergovernmental Panel on Climate Change. Cambridge, United Kingdom and New York, NY, USA: Cambridge University Press; 2014

[4] Sunamura T. The Geomorphology of Rocky Coasts. Chichester: Wiley; 1992. p. 302

[5] Valiela I. Global Coastal Change. Oxford: Blackwell Publishing; 2006. p. 367

[6] Trenhaile AS. Climate change and its impact on rock coasts. Geological Society London Memoirs. 2014;**40**(1):7-17

[7] Bird E. Coastal Geomorphology: An Introduction. 2nd ed. New York: John Wiley & Sons; 2008. p. 436

[8] Bird E. Encyclopedia of the World's Coastal Landforms. Netherland: Springer; 2010. p. 1470

[9] Trenhaile AS. Modeling cohesive clay coast evolution and response to climate change. Marine Geology. 2010;**227**(1-4):11-20

[10] Brown S, Barton ME, Nicholls RJ. The effect of coastal defences on cliff top retreat along the Holderness coastline. Proceedings of the Yorkshire Geological Society. 2012;**59**: 1-13

[11] Castedo R, Murphy W, Lawrence J, Paredes C. A new process–response coastal recession model of soft rock cliffs. Geomorphology. 2012;**177-178**:128-143

[12] Castedo R, de la Vega-Panizo R, Fernández-Hernández M, Paredes C. Measurement of historical cliff-top changes and estimation of future trends using GIS data between Bridlington and Hornsea–Holderness Coast (UK). Geomorphology. 2015;**230**:146-160

[13] Bray MJ, Hooke JM. Prediction of soft-cliff retreat with accelerating sea-level rise. Journal of Coastal Research. 1997;**13**(2):453-467

[14] Lee EM, Clark AR. Investigation and management of soft rock cliffs. DEFRA. London: Thomas Telford; 2002. p. 388

[15] Leatherman SP. Social and economic costs of sea level rise. In: Douglas BC, Kearney MS, Leatherman SP, editors. Sea Level Rise History and Consequences. San Diego, California: Academic Press; 2001. pp. 181-223

[16] Addo KA, Walkden M, Mills JP. Detection, measurement and prediction of shoreline recession in Accra, Ghana. ISPRS Journal of Photogrammetry and Remote Sensing. 2008;**63**:543-558

[17] Stavrou A, Lawrence JA, Mortimore RN, Murphy M. A geotechnical and GIS based method for evaluating risk exposition along coastal cliff environments: A case study of the chalk cliffs of southern England. Natural Hazards and Earth System Sciences. 2011;**11**:2997-3011

[18] Hapke C, Reid D, Richmond B. Rates and trends of coastal change in California and the regional behavior of the beach and cliff system. Journal of Coastal Research. 2009;**25**:603-615

[19] Young AP, Flick RE, O'Reilly WC, Chadwick DB, Crampton WC, Helly JJ. Estimating cliff retreat in southern California considering sea level rise using a sand balance approach. Marine Geology. 2014;**348**:15-26

[20] Marques FMSF, Matildes R, Redweik P. Sea cliff instability susceptibility at regional scale: A statistically based assessment in the southern Algarve, Portugal. Natural Hazards and Earth System Sciences. 2013;**13**(12):1965-2003

[21] Trenhaile AS. Modeling the erosion of cohesive clay coasts. Coastal Engineering. 2009;**56**(1):59-72

[22] Walkden MJA, Hall JW. A predictive mesoscale model of the erosion and profile development of soft rock shores. Coastal Engineering. 2005;**52**(6):535-563

[23] Walkden MJA, Hall JW. A mesoscale predictive model of the evolution and management of a soft-rock coast. Journal of Coastal Research. 2011;**27**(3):529-543

[24] Leatherman SP. Modeling shore response to sea-level rise on sedimentary coasts. Progress in Physical Geography. 1990;**14**:447-464

[25] Ashton A, Murray AB, Arnoult O. Formation of coastline features by large-scale instabilities induced by high-angle waves. Nature. 2001;**414**(6861):296-300

[26] Kamphuis JW. Recession rate of glacial till bluffs. Journal of Waterway, Port, Coastal, and Ocean Engineering. 1987;**113**(1):60-73

[27] Castedo R, Fernández M, Trenhaile A, Paredes C. Modeling cyclic recession of cohesive clay coasts: Effects of wave erosion and bluff stability. Marine Geology. 2013;**335**:162-176

[28] Trenhaile AS. The Geomorphology of Rock Coast. Oxford: Clarendon Press; 1987. p. 384

[29] Teixeira SB. Coastal hazards from slope mass movements: Analysis and management approach on the Barlavento Coast, Algarve, Portugal. Ocean & Coastal Management. 2014;**102**:285-293

[30] Paredes C, Godoy C, Castedo R. Soft-cliff retreat, self-organized critical phenomena in the limit of predictability? Fractals. 2015:**23**(01):1540009

[31] Walkden M, Dickson M. Equilibrium erosion of soft rock shores with a shallow or absent beach under increased sea level rise. Marine Geology. 2008;**251**:75-84

[32] Collins BD, Sitar N. Processes of coastal bluff erosion in weakly lithified sands, Pacifica, California, USA. Geomorphology. 2008;**97**:483-501

[33] Quinn JD, Rosser NJ, Murphy W, Lawrence JA. Identifying the behavioural characteristics of clay using intensive monitoring and geotechnical numerical modelling. Geomorphology. 2010;**120**:107-122

[34] Wolters G, Müller G. Effect of cliff shape on internal stresses and rock slope stability. Journal of Coastal Research. 2008;**24**(1):43-50

[35] Rushton KR. Groundwater Hydrology: Conceptual and Computational Models. Chichester: John Wiley & Sons; 2003. p. 430

[36] Craig RF. Soils Mechanics. 8th ed. London: CRC Press; 2012. p. 584

[37] Wyllie DC, Mah CW. Rock Slope Engineering, Civil and Mining. 4th ed. New York: Spon Press; 2004. p. 431

[38] Budetta P, Galietta G, Santo A. A methodology for the study of the relation between coastal Cliff erosion and the mechanical strength of soils and rock masses. Engineering Geology. 2000;**56**:243-256

[39] USACE. Slope Stability. 2nd ed. Washington, DC: Department of the Army, U.S. Corps of Engineers; 2003. p. 20314.

[40] Sunamura T. A wave tank experiment on the erosional mechanism at a cliff base. Earth Surface Processes and Landforms. 1982;**7**:333-343

[41] Skafel MG, Bishop CT. Flume experiments on the erosion of till shores by waves. Coastal Engineering. 1994;**23**:329-348

[42] Skafel MG. Laboratory measurement of nearshore velocities and erosion of cohesive sediment (till) shorelines. Coastal Engineering. 1995;**24**:343-349

[43] Wilcock PR, Miller DS, Shea RH, Kerkin RT. Frequency of effective wave activity and the recession of coastal bluffs: Calvert Cliffs, Maryland. Journal of Coastal Research. 1998;**14**(1):256-268

[44] Mano A, Suzuki S. Erosion characteristics of sea cliffs on the Fukushima coast. Coastal Engineering Journal. 1999;**41**(1):43-63

[45] Gorokhovich Y, Leiserowiz A. Historical and future coastal changes in Northwest Alaska. Journal of Coastal Research. 2012;**28**:174-186

[46] Bruun P. The Bruun Rule of erosion by sea-level rise: A discussion of large-scale two- and three-dimensional usages. Journal of Coastal Research. 1988;**4**:627-648

[47] Thieler E, Himmelstoss EA, Zichichi JL, Ergul A. The Digital Shoreline Analysis System (DSAS) Version 4.0—An ArcGIS Extension for Calculating Shoreline Change. U.S. Geological Survey, Woods Hole, MA. 2009

[48] Morton RA, Miller TL, Moore LJ. National assessment of shoreline change part 1: Historical shoreline changes and associated coastal land loss along the US Gulf of Mexico. US Geological Survey Open File Report 2004-1043. 2004

[49] Lee EM. Problems associated with the prediction of cliff recession rates for coastal defence. In: Hooke JM, editor. Coastal Defence and Earth Science Conservation. Bath: Geological Society Publishing; 1998. pp. 46-57

[50] Lee EM, Hall JW, Meadowcroft IC. Coastal cliff recession: The use of probabilistic prediction methods. Geomorphology. 2001;**40**:253-269

[51] Dawson DJ, Dickson ME, Nicholls RJ, Hall JW, Walkden MJA, Stansby PK, Mokrech M, Richards J, Zhou J, Milligan J, Jordan A, Pearson S, Rees J, Bates PD, Koukoulas S,

Watkinson AR. Integrated analysis of risks of coastal flooding and cliff erosion under scenarios of long term change. Climatic Change. 2009;**95**(1-2):249-288

[52] Hoek E. Estimating Mohr-Coulomb friction and cohesion values from the Hoek-Brown failure criterion. International Journal of Rock Mechanics and Mining Sciences & Geomechanics Abstracts. 1990;**27**(3):227-229

[53] Kogure T, Aoki H, Maekado A, Hirose T, Matsukura Y. Effect of the development of notches and tension cracks on instability of limestone coastal cliffs in the Ryukyus, Japan. Geomorphology. 2006;**80**:236-244

[54] Kogure T, Matsukura Y. Critical notch depths for failure of coastal limestone cliffs: Case study at Kuro-shima Island, Okinawa, Japan. Earth Surface Processes and Landforms. 2010;**35**(9):1044-1056

[55] Corominas J. The angle of reach as a mobility index for small and large landslides. Canadian Geotechnical Journal. 1996;**33**:260-271

[56] Noetzli J, Huggel C, Hoelzle M, Haeberli W. GIS-based modelling of rock-ice avalanches from Alpine permafrost areas. Computational Geosciences. 2006;**10**:161-178

[57] Berridge NG, Pattinson J. Geology of the Country around Grimsby and Pattrington. London: Her Majesty's Stationery Office; 1994

[58] Catt JA. The pleistocene glaciations of Eastern Yorkshire: A review. Proceedings of the Yorkshire Geological Society. 2007;**56**:177-207

[59] Bell FG. Engineering Properties of Soils and Rocks. 4th ed. Oxford: Blackwell Science Ltd; 2000. p. 492

[60] Bell FG. The geotechnical properties of some till deposits occurring along the coastal areas of eastern England. Engineering Geology. 2002;**63**:49-68

[61] Quinlan PJ. A geomorphological assessment of the stability of the flat cliffs landslide complex, Filley Bay, North Yorkshire [thesis]. University of Leeds; 2005

[62] Lowe JA, Howard TP, Pardaens A, Tinker J, Holt J, Wakelin S, Milne G, Leake J, Wolf J, Horsburgh K, Reeder T, Jenkins G, Ridley J, Dye S, Bradley S. UK Climate Projections Science Report: Marine and Coastal Projections. Exeter: Met Office Hadley Centre; 2009. p. 99

[63] Walkden MJA, Rossington K. Characterisation and prediction of large scale, long-term change of coastal geomorphological behaviours: Proof of concept modeling. Environment Agency; Defra Flood and Coastal Erosion Risk Management Research and Development Programme; 2009

[64] East Riding of Yorkshire Council. Coastal Information Pack [Internet]. 2016. Available from: http://www.eastriding.gov.uk/coastalexplorer/homepage.html. [Accessed: 09 December 2016]

[65] Montreuil AL, Bullard JE. A 150-year record of coastline dynamics within a sediment cell: Eastern England. Geomorphology. 2012;**179**:168-185

Section 3

Trends in Geomorphology

Chapter 5

Digital Elevation Models in Geomorphology

Bartłomiej Szypuła

Additional information is available at the end of the chapter

http://dx.doi.org/10.5772/intechopen.68447

Abstract

This chapter presents place of geomorphometry in contemporary geomorphology. The focus is on discussing digital elevation models (DEMs) that are the primary data source for the analysis. One has described the genesis and definition, main types, data sources and available free global DEMs. Then we focus on landform parameters, starting with primary morphometric parameters, then morphometric indices and at last examples of morphometric tools available in geographic information system (GIS) packages. The last section briefly discusses the landform classification systems which have arisen in recent years.

Keywords: geomorphometry, DEM, DTM, LiDAR, morphometric variables and parameters, landform classification, ArcGIS, SAGA

1. Introduction

Geomorphology, the study of the Earth's physical land-surface features, such as landforms and landscapes and on-going creation and transformation of the Earth's surface, is one of the most important research disciplines in Earth science. The term geomorphology was first used to describe the morphology of the Earth's surface in the end of nineteenth century [1]. Geomorphological studies have focused on the description and classification of landforms (geometric shape, topologic attributes, and internal structure), on the dynamical processes characterizing their evolution and existence and on their relationship to and association with other forms and processes [2].

Geomorphology dates back to sixth to fifth century BC, when Xenophanes of Colophon (580–480 BC) speculated that, as seashells are found on the top of mountains, the surface of the Earth must have risen and fallen, or Herodotus (484–420 BC) thought that the lower part of Egypt was a former marine bay referring to the year-by-year accumulation of river-borne silt in the Nile delta region [3]. Geomorphology as an independent scientific discipline developed

INTECH
open science | open minds

© 2017 The Author(s). Licensee InTech. This chapter is distributed under the terms of the Creative Commons Attribution License (http://creativecommons.org/licenses/by/3.0), which permits unrestricted use, distribution, and reproduction in any medium, provided the original work is properly cited.

in the late nineteenth century [4]. The first modern theory of landscape evolution was the 'geographical cycle' expounded by Davis [5–7]. Another important theory, or rather a variation on Davis' scheme was offered by Penck [8, 9] who stated that according to the Davis model, uplift and planation take place alternately.

Geomorphology in the twentieth century experienced rapid evolution and growth, and many overlapping phases of development can be identified. Contemporary geomorphology combines and consists of many individual fields of science, that is, geology, hydrology, meteorology, cartography, geographic information system (GIS), engineering, biology, archaeology, etc. This complexity and the long tradition associated with its development helped geomorphologist to use many research techniques and measurement tools according to the research requirements. It is often said about the four main research directions, which are historical, dynamic, structural, and climatic, in geomorphology. However, as rightly observed by Migoń [10], each area has its own history, dynamic nature of the processes, and is present in specific geological structures and with the participation of specific climatic conditions.

Herein, we will focus on research approach to landform as the main study subject in geomorphology. In general, over the last 50 years four primary approaches were distinguished in geomorphology.

1. Morphography: It describes physical appearance of landforms and is qualitative approach to the form. It is linked with the direct observation of forms *in situ* which allows specifying the appearance of the form and its morphographic classification (plain, hill, valley, ridge, etc.). These terms do not indicate the way of creation of forms rather only determine their external expressions.

2. Morphogenesis: It focuses on explaining the origin of the forms and determine the mechanisms of their contemporary development. Geomorphologists use different methods to determine the nature of the process in the past and the present form.

3. Morphochronology: It aims to specify age of the forms and the age-relationships between adjacent landforms. Geomorphologists examine both absolute as well as relative age between the forms.

4. Morphometry: It deals with establishing geometric features of the landforms on the basis of measurements. This chapter is just dedicated to morphometry that at the beginning was an element of quantitative geomorphology and later became an independent discipline in the Earth sciences—geomorphometry.

The quantitative phase in geomorphology was developed in 1940–1970 and reflected a broader trend within many of the Earth sciences disciplines towards enhanced use of sophisticated technologies (often derived from military purposes) to measure, describe and analyse the Earth's surface features in number of categories. Horton's publications [11, 12] on stream networks and drainage basin processes are classically identified as the precursor to this quantitative movement. Strahler [13] has limited a method for the quantitative analysis of the forms modelled by flowing water and gravitational movements over a longer period of time. The data were obtained mainly from measurements on detailed topographic maps and then applying statistical and morphometric analysis for various calculations and indicators of the fluvial relief within the drainage basin [14–18], karst [19], or glacial relief [20, 21] were made.

Further development of the geomorphometry resulted in working out of new theoretical and methodological basis. Lustig [22] distinguished two approaches for quantitative analysis and Earth's surface characteristics which are (1) the characteristics of individual forms based largely on field surveys and (2) analysis of the surface area as a whole based on analysis of the map. Then Evans [23] proposed the division of geomorphometry into two forms: general and specific. In general, geomorphometry applies to and describes the continuous land surface. It provides a basis for the quantitative comparison of qualitatively different landscapes and it can adapt to the methods of surface analysis used outside geomorphology. Whereas specific geomorphometry applies to discrete landforms, describes selected relief or landform types as well as their geometry and laws of formation and development. Although such divisions may seem somewhat artificial, since the geometry of individual forms consists of the geometry of the surface as a whole, but they are useful to define more closely the principles of various types of morphometric analysis.

At the end of the 1980s and the beginning of the 1990s, with the personal computer revolution, algorithms have been implemented in many raster-based GIS packages (ArcInfo, MicroStation, MicroDEM, etc.) and it was possible to process digital elevation models (DEMs) over fairly large areas. Then began the new era of geomorphometry with new opportunities to visualize and compute land-surface parameters. Now-a-days, geomorphometry is an important component of terrain analysis and surface modelling including measurements of morphometry of continental ice surfaces, characterizing glacial troughs, mapping sea-floor terrain types, guiding missiles, assessing soil erosion, analysing wildfire propagation, and mapping ecoregions [24, 25]; it is a broad field that is important not only in various aspects of Earth sciences but also in engineering, biology, and medicine [26].

2. Digital elevation models (DEMs)

2.1. History and definition

As Pike [27] noted, a numerical description of the ground surface is helpful in addressing many geomorphological problems. In the definition of the subject of this chapter appears the word 'model', that is, a representation, generally in miniature, to show the construction or appearance of something. Meyer [28] neatly expressed it—reality scaled down and converted to a form which we can comprehend. In this case it is used to represent the original situation—approximation of topography. The term and concept of 'terrain model' was first described by Miller and Laflamme [29] but it did not come into general use until the 1960s and even later because of the technological limitations (computers, processors, etc). They defined digital terrain model (DTM) as just a statistical representation of continuous Earth's surface which consists of many points of known co-ordinates x, y, z in the arbitrary co-ordinate system. In the following years, a number of terms related to the presentation of the Earth's surface by numerical methods, that is, Burrough [30] described DEM as a regular gridded matrix representation of the continuous variation of relief over space. Among many terms, the most common and accepted in geomorphometric and GIS terminologies is digital elevation model (DEM) or digital terrain model (DTM). Some authors make distinctions between them, that is, DEM is 'an ordered array of numbers that represent the spatial distribution of elevations above some arbitrary datum in a landscape'. DTM is 'an ordered array of numbers that represent the spatial distribution of terrain attributes'; therefore DEM is a subset of DTM. Moore et al. [31, 32], Weibel and Heller [33], and Li et al. [34] defined DTM as a digital (numerical) representation of the terrain. Herein both terms will be treated as synonyms but, in fact, the concept of DEM is broader and more universal. In summary, we can say that digital elevation model is the set of digital data describing elevation values of Earth's ground surface (or any other surface) which contains additional information about the character of this surface (i.e. structural lines, break lines, water bodies, etc.) and interpolation algorithm, which is the best for approximation (modelling) of the real topography. A DEM is a complete representation of a land surface which means that heights are available at each point in the area of interest [35].

2.2. Types

Due to geometry and data organization there are several basic types of models: raster (GRID), vector triangulated irregular network (TIN), point and hybrid [34]. Herein we will skip point and hybrid models because they are rarely used and absent in the most popular GIS software packages (i.e. ArcGIS, QGIS, MapInfo, Surfer).

Raster model (GRID) is the most widely used digital height data structure during the last years because of its simplicity [36]. The simplicity consists, in fact, that elevations are stored using a regular square grid that is consistent in each part of the study area. All square grids form regular matrix of heights for which plain co-ordinates (x, y) can be easily calculated due to the regular spacing of the grid points (**Figure 1**). This kind of DEM has many advantages, such as simple elevation matrices that record topological relations between data points implicitly and ease of computer implementation [31, 37]. Moreover, DEM is considerably easier to design land-surface parameters and objects using grids because simpler algorithms can be used; grids have a uniform spatial structure and almost all properties of gridded DEMs are defined by a single characteristic which is cell size; a grid model is more suited to the computer models used in image processing and for printing [35]. For this reason, some DEM software packages accept only grid data.

Raster DEMs also have disadvantages, such as grids present under-sampled topography in areas where the topography is complex, and they over-sample smooth topography; re-projection of a grid is slow and sometimes leads to a loss of accuracy (because the initial grid loses its regular structure in a new projection and so it has to be re-calculated); the different distances between grid centres in cardinal and diagonal directions have a negative impact on the precision of many hydrological modelling [35]; the computed upslope flow paths will tend to have zigzag pattern across the landscape and increase the difficulty of calculating specific catchment areas accurately [38, 31]; square grids cannot handle abrupt changes in elevation easily and they will often skip important details of the land surface in flat areas [39].

The second most popular and widely used is a vector model named triangulated irregular network (TIN). TIN is proposed by Hormann [40] which devised a TIN idea, linking selected

Figure 1. Example of the GRID model (A), and TIN model (B).

points on divides, drainage lines and breaks in slope to interrelate height, slope gradient and aspect. TIN is based on triangular elements (facets that vary in shape and size) with vertices at the sample height points [31]. These facets consist of planes joining the three adjacent points in the network and are usually constructed using Delaunay triangulation [33]. TIN can be interpolated directly from surveyed points or discrete features that are extracted manually from maps or by computer from a grid or contour DEM [27]. The triangle may be regarded as the most basic and universal unit in all geometrical patterns (because of their great flexibility in terms of shape and size), since a regular grid of square or rectangular cells or any polygon with any shape can be decomposed into a series of triangles [34].

The advantages of such a model are: relevance to gravitational movements, and especially to hydrological applications [41]; TINs are widely used in perspective representations of surfaces, especially in dynamic fly-through displays where the foreground is represented in full detail but the background can be simplified as larger triangles; for areas with high relief or rougher surfaces, irregular DEMs can use smaller spacing between points, and larger spacing where relief is lower or where the surface is smoother and in this way they can more accurately describe geological faults and other sharp elevation changes using the same number of points as grids [35]; the TIN structure gives best reflecting processes of erosion and deposition mimics paths of steepest gradient [41]. Of course we can use the algorithms to convert the TIN into grid and vice versa, but usually at the loss on the quality and accuracy of the model.

2.3. Data sources

DEM source means the data acquired by different techniques and from different sources. In general, such data can be acquired from topo maps and from field surveys.

Manual or semi-automated digitizing of contour lines and every single elevation point on topographic maps on the computer screen or digitizer can be done relatively cheaper without the need for resurveying. Derived DEMs represent the underlying terrain without surface vegetation and buildings. Despite many critical voices regarding the use of this method [42, 43], it is a quick and effective method to get quite good DEM and till the end of 1990s this was the most common source of elevation data for DEMs.

The second group of data sources are surveys, such as ground survey techniques, digital photogrammetry, and remote sensing. Field surveys are carried out using total stations, theodolite, levelling instruments, global positioning system (GPS), and differential global positioning system (DGPS). These kinds of surveys have high accuracy (sometimes even less than 1 cm), flexibility (the measurement density can be varied, depending on the terrain), and very little processing is required after the measurements have been taken, but especially suited for measuring small areas. Digital photogrammetry relies on the stereoscopic interpretation of aerial photographs or satellite imagery using manual or automatic stereoplotters [39, 33]. Remote sensing surveys consist of Airborne laser scanning (ALS) or satellite platforms and include laser-ranging altimetry (LiDAR—light detection and ranging), synthetic aperture radar interferometry (InSAR or IfSAR). The final results will often include tree-top canopies and buildings. This gives higher elevation values, rough surfaces, and high slope values [43]. The advantages of the LiDAR method are production time that is typically shorter than that for photogrammetrically generated DEMs [44] and great spatial and vertical accuracy (<0.5 m). The main disadvantage of LiDAR data is point cloud which produces a very dense and detailed land-surface model that could be difficult to handle during the production process and sometimes the accuracy of the readings vary according to the characteristics of the terrain (i.e. very steep slopes) [45]. However, these days LiDAR is definitely the best method of the DEM production. Several countries have already produced national LiDAR DEM/DSM (e.g. Belgium, the Netherlands at resolutions of 2–5 m, Poland at 1 m).

2.4. Free global DEMs

Currently, many elevation data with better accuracy and parameters are freely available in the world including LiDAR data. Individual countries or institutions offer different kinds of models. In **Table 1**, there are models (and their properties) available for the whole world, completely free of charge.

Model	Spatial resolution (grid size)	Accuracy vertical/ horizontal	Co-ordinate System	Data format	Data extent	Source	Institution, publication
GTOPO30	30 × 30″ (~1000 m)	10–300/150 m	WGS-84	DTED, USGS DEM, DCW	90°N to 90°S	https://lta.cr.usgs.gov/GTOPO30	USGS (USA) and NASA, UNEP/GRID, USAID, INEGI (Mexico), GSI (Japan), MWLR (New Zealand), SCAR 1996
GLOBE DEM	30 × 30″ (~1000 m)	250/160 m	WGS-84	DTED	90°N to 90°S	https://www.ngdc.noaa.gov/mgg /topo/gltiles.html	NOAA, 1999
SRTM v3	1 × 1″ (~30 m) 3″ × 3″ (~90 m)	10/13 m	WGS-84/ EGM96	DTED, BIL, GeoTIFF, hgt	60°N to 54°S	https://earthexplorer.usgs.gov/	NASA, NGA, 2013
ETOPO1	1 × 1″ (~30 m)	?	WGS-84	netCDF, GRD98, xyz, GeoTIFF	90°N to 90°S	https://www.ngdc.noaa.gov/mgg/global/global.html	NOAA, 2009
ASTGTM v02 **AST14DEM v03**	1 × 1″ (~30 m) 1 × 1″ (~30 m)	20/30 m	WGS-84/ EGM96 WGS-84/UTM	GeoTIFF	83°N to 83°S	https://search.earthdata.nasa.gov/search?q=ASTGTM+V002	NASA, METI, 2009, 2011
GMTED2010	7.5′ × 7.5″ (~250 m) 15 × 15′ (~500 m) 30″ × 30′ (~1000 m)	26–30 m/? 29–32 m/? 25–42 m/?	WGS-84	ESRI ArcGrid, GeoTIFF	84°N to 56°S	https://topotools.cr.usgs.gov/gmted_viewer/viewer.htm	USGS, NGA (USA), 2011
AW3D30	1 × 1″ (~30 m)	5/5 m	ITRF97/GRS80	GeoTIFF	82°N to 82°S	http://www.eorc.jaxa.jp/ALOS/en/aw3d30/data/index.htm	JAXA, 2016

Table 1. Basic information about the free global DEMs.

GTOPO30 (Global 30 arc-second Elevation): This is a global DEM with a horizontal grid spacing of 30 arc seconds (approx. 1 km) resulting in a product having dimension of 21,600 rows and 43,200 columns. GTOPO30 was derived from eight several raster and vector sources of topographic information which are digital terrain elevation data (DTED), Digital Chart of the World, USGS 1-degree DEM's, army map service (AMS) 1:1,000,000, International Map of the World 1:1,000,000, Peru Map 1:1,000,000, New Zealand DEM, and Antarctic Digital Database [46]. The horizontal accuracy is ±150 m linear error at 90% confidence and vertical accuracy is 10–300 m at the 90% confidence level [47–49].

GLOBE DEM (Global Land 1-km base elevation DEM): It is one of the first global DEM, which was initially spearheaded in 1990. This data set covers 180°W to 180°E longitude and 90°N to 90°S latitude. The horizontal resolution is 0.5 arc-minute in latitude and longitude, resulting in dimensions of 21,600 rows and 43,200 columns [50]. GLOBE version 1.0 has 11 broad sources of information: 6 gridded DEMs and 5 cartographic sources, which were adapted for use in GLOBE. These were digital terrain elevation data (DTED), Digital Chart of the World, Australian DEM, Antarctic Digital Database, Brazil Maps 1:1,000,000, DEM for Greenland, army map service (AMS) Maps 1:1,000,000, DEM for Japan, DEM for Italy, DEM for New Zealand, and Peru Map 1:1,000,000 [51]. Vertical accuracy expressed as ±30 m linear error at 90% confidence can also be described as a root mean square error (RMSE) of 18 m. Linear error distribution is 97 m for RMSE and it can be expressed as ±160 m linear error at 90% confidence.

Shuttle radar topography mission v3 (SRTM): It was flown to aboard the space shuttle **Endeavour** February 11–22, 2000. The National Aeronautics and Space Administration (NASA) and the National geospatial-intelligence agency (NGA) participated in an international project to acquire radar data that was used to create near-global set of land elevations. SRTM successfully collected radar data over 80% of the Earth's land surface between 60°N and 56°S latitude with data points posted every 1 arc-second (approximately 30 m). Absolute height error of SRTM data sets is 5–10 m and absolute geolocation error is 7–13 m [52]. The level of processing and the resolution of the data vary by SRTM data set:

1. **SRTM Non-Void Filled**. This version was edited or finalized by the NGA to delineate and flatten water bodies, better define coastlines, remove spikes and wells, and fill small voids. Data were sampled at 1 arc-seconds (USA) and 3 arc-seconds (rest of world);

2. **SRTM Void Filled**. This elevation data are the result of additional processing to address areas of missing data or voids in the SRTM Non-Void Filled collection. The resolution for SRTM is 1 arc-seconds (USA) and 3 arc-seconds (rest of world);

3. **SRTM 1 Arc-Second Global**. This elevation data offer worldwide coverage of void filled data at 1 arc-second (30 m) resolution and provides open distribution of this high-resolution global data set. Some tiles may still contain voids. Please note that tiles above 50°N and below 50°S latitude are sampled at a resolution of 2 by 1 arc-second [53].

ETOPO1 (Global Relief Model at 1 arc-minute resolution): It was made in 2008 by the National Geophysical Data Centre (NGDC), an office of the National Oceanic and Atmospheric Administration (NOAA). This model was developed as an improvement to the ETOPO2v2

Global Relief Model. ETOPO1 is available in two versions: 'Ice Surface' (top of Antarctic and Greenland ice sheets) and 'Bedrock' (base of the ice sheets). These versions of ETOPO1 were generated from diverse (regional and global) digital data sets, which were shifted to common horizontal and vertical datum. Next steps were evaluated and edited as needed [54]. Shoreline, bathymetric, topographic, integrated bathymetric–topographic, and bedrock digital data sets (13 sources) were obtained from several U.S. government agencies, international agencies, and academic institutions.

ASTGTM (ASTER global digital elevation model v002 and AST14DEM—ASTER digital elevation model v003): These DEMs were developed jointly by the US NASA and Japan's Ministry of Economy, Trade and Industry (METI). ASTER is capable of collecting in-track stereo using nadir and aft-looking near infrared cameras. Since 2001, these stereo pairs have been used to produce single-scene (60 × 60 km) DEMs having vertical RMSE accuracies generally between 10 and 25 m [55]. The ASTER GDEM covers land surfaces between 83°N and 83°S and is comprised of 22,702 tiles. This model is distributed as georeferenced tagged image file format (GeoTIFF) files. The data have resolution at 1 arc-second (approximately 30 m) grid and referenced to the 1984 World Geodetic System (WGS84)/1996 Earth Gravitational Model (EGM96) geoid. Although the ASTER GDEM v. 002 is better model than ASTER GDEM v. 001, users have to know that the data still may contains anomalies and artefacts. One should know that these mistakes can introduce large elevation errors on local scales [56].

GMTED2010 (Global Multi-resolution Terrain Elevation Data 2010): This model was made by collaborating USGS and the NGA, which replaced GTOPO30 model. The new elevation data set has been generated at three separate horizontal resolutions of 30 (about 1 km), 15 (about 500 m), and 7.5 arc-seconds (about 250 m). The global aggregated vertical accuracy of GMTED2010 can be summarized in root mean square error (RMSE). At 30 arc-seconds, the RMSE range is 25–42 m; at 15 arc-seconds, range is 29–32 m, and at 7.5 arc-seconds, range is in between 26 and 30 m. This new product suite provides global coverage of all land areas from latitude 84°N to 56°S for most products, and coverage from 84°N to 90°S for several products [57]. An additional advantage of the new multi-resolution model over GTOPO30 is that at each resolution seven new raster elevation data sets are available. The new models have been produced using the various aggregation methods: minimum elevation, maximum elevation, mean elevation, median elevation, standard deviation of elevation, systematic subsample, and breakline emphasis. GMTED2010 is based on data derived from different raster-based elevation sources, such as SRTM DTED, non-SRTM DTED, Canadian digital elevation data (CDED), Satellite Pour l'Observation de la Terre (SPOT 5), Reference3D, National elevation dataset (NED), GEODATA, an Antarctica satellite radar and laser altimeter DEM, and Greenland satellite radar altimeter DEM [57].

AW3D30 (ALOS Global Digital Surface Model): The global digital surface model (DSM) at 1 arc-second (approximately 30 m) resolution that was released by Japan aerospace exploration agency (JAXA). This model has been compiled with images acquired by the advanced land observing satellite (ALOS). The elevation data are published based on the DSM data set (5-m mesh version) of the 'World 3D Topographic Data' [58]. The source data were a huge amount of stereo-pairs images derived from satellite mission in the years 2006–2011. Next, they were processed semi-automatically to provide digital surface model (DSM). The height accuracy of the data set is approximately <5 m from the evaluation with ground control points (GCPs) or reference DSMs derived from the LiDAR [59].

In addition to the above global DEMs, there are many elevation data with much better accuracy, but for smaller areas, for example, for Europe EU-DEM with grid size 25 m, for Spain MDT05/MDT05-LIDAR (5 m), for Netherlands AHN3 DTM (0.5 m), for all Slovenia (LiDAR data), or LiDAR data for Haiti (1 m).

3. Morphometric landform properties

The main object of the studies in geomorphology is the relief (of the Earth's surface). The term relief of the Earth's surface was used to describe the vertical dimension or amplitude of topography [60] or as the elevation difference over a pre-determined area or collective elevations and their inequalities of a land surface [61]. One can also say that the relief is the complexity of the Earth's surface shapes, and these shapes form the landforms. If we would present a continuous Earth's surface as a matrix of discrete points with a defined distance interval from one another, the shape dimensions would express in the changes of height, distance, and direction between these points. These three vectors and the relationships between them are the basis for most morphometric landform properties.

3.1. Morphometric parameters

Elementary data used in geomorphometry are DEM. Strength of DEM is a suggestive (plastic and three-dimensional) visual message and the ability of quantifying the topography. Quantification of the Earth's surface is expressed by topographic attributes which can be determined from the derivatives of the topographic surface. Generally, one can say, these derivatives measure rate at which elevation changes in response to changes in location. Deng et al. [62] noted that local terrain shape (as the continuous variation of elevation values over the terrain surface from point to point) has an enormous impact on local terrain attributes, but this role is influenced by data and computational factors.

Theoretical assumptions concerning the morphometric properties of the surface come from the 1950s [63–65], and even earlier. Stone and Dugundji [66] and Hobson [67] stated that measures of landforms can be considered as a kind of the roughness of the surface. In general, roughness refers to the irregularity of a topographic surface and cannot be completely defined by any single measure but must be represented by a roughness vector or set of parameters. With roughness, concept of wavelength and amplitude ideas was related. The significant wavelengths of topography are termed grain or texture, while amplitudes associated with these wavelengths correspond to the concept of relief [60]. Texture and grain are terms that have been used to indicate in some way the scale of horizontal variations in the topography. Texture is used to refer the shortest significant wavelength in the topography and grain is used for the longest significant wavelength. Texture is related to the smallest landform elements, while the grain is related to the size of area over which one measures other parameters.

Wood and Snell [68] defined grain as the size of area over which the other factors are to be measured. It is dependent on the spacing of major ridges and valleys and thus indicates texture of topography. Texture refers to the shortest significant topographic wavelength.

The systems proposed by Evans [23] and Krcho [69] for field variables are local-based, relating to a vanishingly small area around each point. They divided topographic properties into primary geomorphometric parameters (height, slope, aspect, profile, and planar curvature) and statistical measures derived in square-matrices out of the gridded DEMs (such as relief, standard deviation, and skewness of heights).

Mark [60] stated, that probably the most important single class of processes which has shaped the Earth's surface could be divided into geometrical properties that involve the relationships among dimensional properties, such as elevation, lengths, areas, volumes, and topological properties which relate numbers of objects in the drainage network (e.g. the bifurcation ratio). Then Moore and Thornes [70] developed LEAP-land erosion analysis programs to examine the spatial distribution of slope length, slope steepness, and the plan curvature for assessing topographic erosion potential.

Pike [71] noted that terrain can be abstracted using geometric signature—a set of measurements that describes topographic form sufficiently well to distinguish geomorphically disparate landscapes. Then he developed the concept of a geometric signature, a multi-variate description of topography using a suite of measures, and later expanded the concept with a listing of 49 variables that could be grouped into 22 attributes [72, 73]. He considered roughness and height, the two most important attributes, with two measures of texture at seventh and eleventh position. Fifteen different variables contribute to roughness. Pike et al. [74] also referred to grain concept. Relief at topographic grain is an estimate of local relief optimized by varying unit-cell size. In homogeneous terrain, local relief within nested circles increases with circle size and then levels off at a diameter termed grain, a measure of characteristic local ridgeline-to-channel spacing. Topographic grain is the characteristic horizontal spacing of major ridges and valleys and is an important descriptor of meso-scale topographic texture. The grain concept arose from the need for a variable and non-arbitrary unit-cell size and also enables to calculate local relief parameter.

Moore and Grayson [41] and Moore et al. [31, 32] described terrain attributes adapted from Speight [75, 76], and then often mentioned and cited [3, 36, 77–80]. They divided terrain attributes into primary and secondary (compound) attributes. Primary attributes, hydrologically related, are directly calculated from elevation data and include variables, such as altitude, upslope height, aspect, slope, upslope slope, dispersal slope, catchment slope, upslope area, dispersal area, catchment area, specific catchment area, flow path length, upslope length, dispersal length, catchment length, profile curvature, and plan curvature. It is interesting that Bork and Rohdenburg [81] have developed a digital relief model for estimating and displaying the distribution of these morphographic parameters.

Schmidt and Dikau [78] subdivided primary geomorphometric parameters into three types: simple, complex, and combined. Simple primary geomorphometric parameters (usually derived through a filter operation within a 3 × 3 moving window) are height, slope, aspect, profile curvature, contour curvature, drainage direction, and real area of pixel. Complex geomorphometric parameters are derived through the analysis of the whole matrix of a DEM. They contain structural information about the surrounding morphometry and consist of contributing area, mean slope of contributing area, average, and variance of primary parameters in contributing area, length of flowpath to outlet, average and variance of primary parameters in flowpath to outlet, length of flowpath to stream, average and variance of primary parameters in flowpath to stream, x- and y-co-ordinates of corresponding stream point, height of corresponding stream point, height distance to corresponding stream point, length of minimum flowpath to watershed, relative slope position after minimum, slope length, relative slope position after maximum, and minimum and maximum slope length.

Interesting approach to the morphometric parameters was presented by Shary et al. [82], who stated that morphometric variables describe not the land surface itself, but rather the system, land surface + vector field, where vector fields of common interest are gravitational field and solar irradiation. He divided morphometric variables and concepts into field-specific (may refer to this system description) and field-invariant (invariant with respect to any vector field, that is, describing the land surface geometrical form). On the other hand, morphometric variables may divide into local, regional, or global (when height data of all the Earth is needed for their determination).

Basso [79] divided topographic attributes into: (i) local (calculated from a small neighbouring area surrounding the DEM cell, usually 3 × 3), (ii) regional (calculated using considerably larger geometric area than the local attributes, less sensitive to the DEM resolution), (iii) catchment oriented (related to the whole catchment area, and are the measurement of certain catchment characteristics), and (iv) process oriented (describe or characterize the spatial variability of a simple representation of specific processes that occur on the landscape).

Goodwin and Tarboton [84] described five categories of the morphometric parameters of drainage basin which are size properties (provide measures of scale that can be used to compare the magnitudes of two or more drainage basins), surface properties (quantities depicted by fields comprising a value at each point within a domain, drainage basin), shape properties (i.e. length, width, perimeter, and more complex function), relief properties (total basin relief, relief ratio [85], and hypsometric curve), and texture properties (amount of landscape dissection by a channel network).

Olaya [86] presented division of the morphometric parameters into local and regional. Local parameters consist of geometric (slope, aspect, curvatures, visibility, visual exposure, and visibility index) and statistical (i.e. average value, standard deviation, skewness coefficient, kurtosis coefficient, range of values, etc.). Regional parameters are connected with hydrological properties (catchment area, height, slope, proximity, etc. [86]).

The latest approach to the classification system of the fundamental geomorphometric variables is presented by Evans and Minár [87]. They proposed field and object variables. Field variables include specific to gravity field (local point-based, local area-based, and regional), specific to other fields and field-invariant variables (local point- and regional-based). Object variables differ between areal, linear, and point features.

Thus the optimal number of variables depends on spatial scale, resolution of the source data (DEM), and the requirements of the research problem; and many measures describe the same attribute of surface form and thus are redundant; for this reason new geomorphometric parameters are very rare.

3.2. Morphometric indices

Morphometric parameters, which were discussed in the previous section, showed that there are many different classifications of these parameters. In this section, we want to look at some popular morphometric indices called secondary or composite attributes [3, 32, 36, 62, 80, 89], combined or compound geomorphometric parameters [78, 83], statistical parameters [23, 86] or process oriented [79]. These indices are combinations of the primary morphometric attributes and describe or characterize the spatial variability of specific processes occurring on the landscape. Sometimes these morphometric indices can be derived empirically but it is preferable to develop them through the application and simplification of the underlying physics of the processes. With the index approach we simplify some physical sophistication to allow improved estimates of spatial patterns in the landscape [31].

In **Table 2**, some selected geomorphometric indices commonly used and their definitions were listed.

Parameter	Formula	Description	Source
Drainage density	$D_D = L/A$ where: L is sum of the channel lengths and A is basin area	The sum of the channel lengths divided by basin area. It is important indicator of the linear scale of landform elements in a drainage basin, indicates the closeness of spacing of channels, thus providing a quantitative measure of the average length of stream channel for the whole basin.	Horton [12]
Form factor	$R_f = A(lb)^2$ where: A is area of basin, Lb is the basin length	The ratio of the basin area to the square of the basin length. Indicates the flow intensity of a basin of a defined area. The form factor value should be always less than 0.7854 (the value corresponding to a perfectly circular basin).	Horton [12]
Terrain ruggedness index (TRI)	$\mathrm{TRI} = Y[\Sigma(x_{ij}-x_{00})^2]^{1/2}$ where: x_{ij} is elevation of neighbour cell to cell (0,0)	Measure of topographic heterogeneity (amount of elevation difference between adjacent cells of a DEM).	Riley et al. [90]
Stream power index	$\Omega = pgq \cdot \tan\beta)$ where: pg is the unit weight of water, q is the discharge per unit width and β is the representative slope angle	This is the time rate of energy expenditure and has been used extensively in studies of erosion, sediment transport and geomorphology as a measure of the erosive power of flowing water.	Moore et al. [31]
Elevation-relief ratio	$E = (H_{mean} - H_{min})/(H_{max} - H_{min})$ where: H is elevation	Expresses relative proportion of upland to lowland within a sample region. Usually ranges from 0.15 to 0.85. Low values occur in terrains characterized by isolated relief features standing above extensive level surfaces, and high values describe broad and level surfaces, broken by occasional depressions.	Pike and Wilson [91] after Wood and Snell [68]

Parameter	Formula	Description	Source
Topographic openness algorithm	Positive openness $\phi L = (0\phi L + 45\phi L + \ldots + 315\phi L)/8$ negative openness $\Psi L = (0\Psi L + 45\Psi L + \ldots + 315\Psi L)/8$ where: L is specified distance	Describes the degree of enclosure of a location on an irregular Earth's surface. Positive values expressing openness over surface topography is high for convex forms, whereas negative values expressing and describing attribute below the surface topography and are high concave forms.	Yokoyama et al. [92]
Terrain shape index	$TSI = \bar{Z}/R$ where: Z is mean elevation of the sample plot boundary, and R is plot radius measured in the units used for elevation	This index is equivalent to the mean slope gradient of the plot boundary as viewed from the plot centre, with units of meters change in elevation per meter of plot radius. Typical TSI values for mountain landforms ranged from −0.24 to −0.12, on convex upper slopes near ridge tops, and from +0.09 to +0.17 on concave lower slopes.	McNab [93]
Landform index	$LI = H^{\circ}/100$ where: H° is vertical gradient to the horizon	The landform index is the average vertical gradient to the topographic horizon, divided by 100 to convert percent to a decimal value. The index is dimensionless and the effects of height and distance to the landform are compensating factors.	McNab [94]
Compound topographic index (CTI)	$CTI = \ln (A_f/\tan\beta)$ where: A_f is the specific catchment area draining through the point, and β is the representative slope angle	Ratio between slope and catchment area; quantification of catenary topographic convergence represented by slope angle and catchment. For the same contributing area CTI values are higher for pixels with lower slopes —this means that CTI primarily reflects accumulation processes.	Moore et al. [32]
Heat load index	$HLI = [1 - \cos(\theta - 45)]/2$ where: θ is aspect in degrees east of north	Quantitative measure of aspect and steepness of slope. The equations applied to 0–60° north latitude, slopes from 0 to 90° and all aspects.	McCune and Keon [95]
Anisotropy index	$ANI = \hat{R}_{min}/\hat{R}_{max}$ where: $\hat{R}_{min}$ is smallest estimated range parameter, and $\hat{R}_{max}$ is highest estimated range parameter in various directions	Ratio between the minimum and maximum range parameter of spatial dependence, fitted for various directions	Bishop and Minasny [96]
Shape complexity index	$SCI = P/(2r\pi), r = \sqrt{\frac{A}{\pi}}$ where: P is the perimeter of the polygon, A its area, r is the radius of the circle with the same area	Index which is used to describe polygons on DEM slices. Indicates how compact (or oval) a feature is.	Hengl et al. [97]
Basin relief ratio	$Rh = H/L$ where: H is total basin relief, and L is basin length	Ratio between total basin relief (difference in elevation of basin mouth and summit) and basin length, measured as the longest dimension of the drainage basin. Indicates overall slope of the watershed surface. It is a dimensionless number, readily correlated with other measures that do not depend on total drainage basin dimensions.	Schumm [98]

Parameter	Formula	Description	Source
Relative relief	**Rp** = H/P where H is total basin relief, and P is basin perimeter	Ratio between total basin relief and drainage basin perimeter.	Melton [99, 100]
Drainage basin compactness	$B_c = P/A$ where: P is drainage basin perimeter, and A is drainage basin area	Ratio between perimeter and area of drainage basin. Higher values correspond to the basins of developing the long-term share erosion running in conditions of relative peace tectonic or are typical for catchment formed in low resistance rocks.	Engstrom [101]
Drainage basin shape ratio	$B_s = B_l/B_w$ where: B_l is max length of the drainage basin, and B_w is max width of the drainage basin	Ratio between maximum length and maximum width of drainage basin. Higher values correspond to more elongated basins and also indicate a relatively higher tectonic activity of the area.	Cannon [102], Ramírez-Herrera [103]
Surface area ratio	**SAR** = A/A_p where: A is cell's surface area, and A_p is planimetric area of that cell	Ratio between surface area and planimetric area; surface ratios will always be greater than or equal to 1.	Jennes [104]
Valley height-width ratio	$V_f = 2V_{fw}/[(E_{ld} - E_{sc}) + (E_{rd} - E_{sc})]$ where: V_{fw} is width of the valley floor, E_{ld}/E_{rd} height of the left/right watershed and E_{sc} height of the valley floor	Ratio of the width of the valley bottom and the average height of the slopes; allows comparison of erosional patterns between watersheds. Low index value V_f characteristics are deeply cut river valleys developing under high uplift area, while high values correspond to the valleys in the wide areas of thin tectonic activity.	Bull and McFadden [105]
Dissection index	**DI** = R_R/A_R where: R_R is relative relief, and A_R is absolute relief	Ratio between relative and absolute relief which always varies between zero (complete absence of dissection) and one (extreme case, vertical cliff). Assess the degree of incision of a landscape.	Sharma [106]
Relief index (RI)	**RI** = C_L/A_P where: C_L is total length of contour lines, and A_P is planar surface area	Ratio of the summary length of the contour lines and the planar surface area at which they occur; shows relief variability. RI is based on a combination of local relief (number of contour lines and elevational changes) and degree of surface cut (length and shape of the contour lines) with reference to the planar surface area.	Szypuła [107]

Table 2. Selected geomorphometric indices derived from DEMs.

3.3. Morphometric tools available in GIS packages

The current availability of high speed computing platforms and high-resolution (less than 10 m spatial resolution) DEMs now provides the opportunity to perform quantitative analyses and calculating morphometric indices on new level. Many GIS packages offer tools to work with DEMs. On one hand we have the comWmercial software (ArcGIS, MapInfo, Surfer, Global

Mapper, Terra Solid), and on other one, great free programs (SAGA, QGIS, MicroDEM), which offer a lot of interesting tools. In **Table 3**, we have presented useful tools for geomorphometric (or geomorphological) analysis. As examples we chose one commercial package (ArcGIS) and one free (SAGA). One should remember that rapid change of the GIS applications cause, next new versions a large number of greater tools.

SAGA 3.0.0	ArcGIS 10.5
Terrain analysis—Channels	**3D analyst tools—Raster Surface**
Channel network	Aspect
Channel network and drainage basins	Contour
Overland flow distance to channel network	Contour list
Strahler order	Contour with barriers
Valley depth	Curvature
Vertical distance to channel network	Cut fill
Watershed basins	Hillshade
Watershed basins (Extended)	Slope
Terrain analysis—compound analyses	**3D analyst tools—visibility**
Elevation	An overview of the visibility toolset
Analytical hillshading	Construct sight lines
Slope	Intervisibility
Aspect	Line of sight
Plan curvature	Observer points
Profile curvature	Skyline
Convergence index	Skyline barrier
Closed depressions	Skyline graph
Total catchment area	Sun shadow volume
Topographic wetness index	Viewshed
LS-factor	Viewshed 2
Channel network	Visibility
Drainage basins	**Spatial analyst tools—solar radiation**
Channel network base level	Area solar radiation
Channel network distance	Points solar radiation
Valley depth	Solar radiation graphics
Relative slope position	* Exemplary additional installed tools :

SAGA 3.0.0	ArcGIS 10.5
Terrain analysis—hydrology	**Hypsometry tools**
Cell balance	Hypsometric curve
Downslope area (interactive)	Area gradient tools
Edge contamination	Max watershed
Flow accumulation (Flow Tracing)	**Surface area analysis**
Flow accumulation (Mass-Flux Method)	Surface area
Flow accumulation (Recursive)	Surface ratio
Flow accumulation (Top-Down)	Flat area
Flow depth (interactive)	**Geomorphometric and gradient metrics**
Flow path length	Classify aspect
Flow sinuosity (interactive)	Linear aspect
Flow width and specific catchment area	Mean slope
Isochrones constant speed (interactive)	Slope impedance
Isochrones variable speed (interactive)	Dissection
LS factor	Hierarchical slope position
LS-factor, field-based	Landform curvature
Lake flood	Roughness
Lake flood (interactive)	Slope position
Melton ruggedness number	Surface/area ratio
SAGA wetness index	Surface relief ratio
Slope length	Compound topographic index
Slope limited flow accumulation	Heat load index
Stream power index	Integrated moisture index
TCI low	Site exposure index
Topographic wetness index (TWI)	Topographic radiation aspect index
Upslope area	
Upslope area (interactive)	
Terrain analysis—lighting, visibility	**Topography tools**
Analytical hillshading	Beer's aspect
Potential incoming solar radiation	McCune and Keon heat load index
Sky view factor	Landform classification
Topographic correction	PRISM data helper
Topographic openness	Slope position classification
Visibility (points)	Solar illumination index
Visibility (single point) (interactive)	Topographic convergence/wetness index
	Topographic position index

SAGA 3.0.0	ArcGIS 10.5
Terrain analysis—morphometry	**Local relief model**
Convergence index	**Terrain tools**
Convergence index (Search radius)	Terrain ruggedness (VRM)
Curvature classification	**Riparian topography tools**
Diurnal anisotropic heating	Calculate flooding height
Downslope distance gradient	Calculate inundation area
Effective air flow heights	Height above river
Fuzzy landform element classification	Prepare HAR for flooding
Hypsometry	**Flow accumulation for both positive and negative values**
Land-surface temperature	**Land facet analysis**
Mass balance index	Calculate density raster
Morphometric features	Shannon's diversity index
Morphometric protection index	Identify termini polygons
Multi-scale topographic position index (TPI)	Topographic position index tools
Multi-resolution index of valley bottom flatness (MRVBF)	Mahalanobis distance tools
Real surface area	
Relative heights and slope positions	
Slope, aspect, curvature	
Surface specific points	
TPI-based landform classification	
Terrain ruggedness index (TRI)	
Terrain surface classification (Iwahashi and Pike)	
Terrain surface convexity	
Terrain surface texture	
Topographic position index (TPI)	
Upslope and downslope curvature	
Valley and ridge detection (Top hat approach)	
Vector ruggedness measure (VRM)	
Wind effect (Windward/Leeward Index)	
Wind exposition index	

Table 3. The list of exemplary morphometric tools available in SAGA and ArcGIS software (after [108–110]).

4. Landform classifications

Geomorphology studies the relief. If we want to understand the relief of the Earth's surface (which is highly complex) we need to simplify and subdivide it into landforms. We have to focus on describing landforms, their spatial arrangement and the processes which led to their formation. Of course, many landforms can be delineated manually using photo-interpretation to assess their form, size, scale, adjacency, surface roughness, hydrological and contextual position but there is always problem with the boundary of the landform. And herein DEMs are helpful. Availability of global DEMs and high accuracy national LiDAR data made new possibilities of analysing these data to extract and classify geomorphic entities. The landform elements can be extracted automatically from DEMs by using land-surface parameters, such as slope, curvatures, catchment area, distance to streams, peaks and depression depth, etc. [111]. The goal of automated extraction of landforms and landform elements using semi-automated or fully-automated algorithms is to find their geometric signature, which Pike [72] defined as a set of measures that describe topographic form well enough to distinguish among geomorphically disparate landscapes.

There are many landform classification systems, and herein we take this issue very briefly and in general. The pioneer works in quantitative systems for landform classification were conducted by Hammond [112], Wood and Snell [68] and Anstey [113, 114] by using topographic contour maps. These studies were aimed primarily towards the systematization and logical interpretation of terrain data to assist in the determination of design criteria for materiel, and secondarily towards the development of a universal system for the quantification of landform data.

In the mid-1960s, Hammond [115] devised the three-level system of regional landform classification, based purely on geomorphometric parameters calculated in the chosen window size (**Figure 2**). For each window position the following parameters were calculated: (1) percentage of area where the ground is flat or gentle (less than 8% slope); (2) local relief (maximum minus minimum elevation); and (3) profile type (relative proportion of flat or gently sloping terrain that occurs in lowlands or uplands).

Peucker and Douglas [116] showed a few methods designed to detect pits, peaks, passes, ridges, ravines, and breaks, given an array of sampled, quantized terrain heights. Described methods used local analysis which means the results at each point do not depend on the results already achieved at other points.

As the hillslopes constitute a basic element of all landscapes and a fundamental component of geomorphologic systems, many subjective classifications of slope profiles that were intended to create conceptual classifications of hillslopes have been proposed. Ruhe [117] divided hillslopes into summits, shoulders, backslopes, footslopes, and toeslopes. Dalrymple et al. [118] and Conacher and Dalrymple [119] proposed nine unit classification of hillslopes: (i) interfluve, (ii) seepage slope, (iii) convex creep slope, (iv) fall face, (v) transportational midslope, (vi) colluvial footslope, (vii) alluvial toeslope, (viii) channel wall, and (ix) channel bed (after [120]).

Next, there were several approaches to standardize hillslope units using qualitative terms [121]. Most commonly, a hillslope was described by a series of basic units describing changes in slope, curvature, and processes along the hillslope profile [122].

As Young [123] noted the curvature can be classified into convex, concave, and rectilinear surfaces, Dikau [124] defined basic form elements of the landscape as the combination of three slope profile curvature characteristics and three plan curvature characteristics which lead to

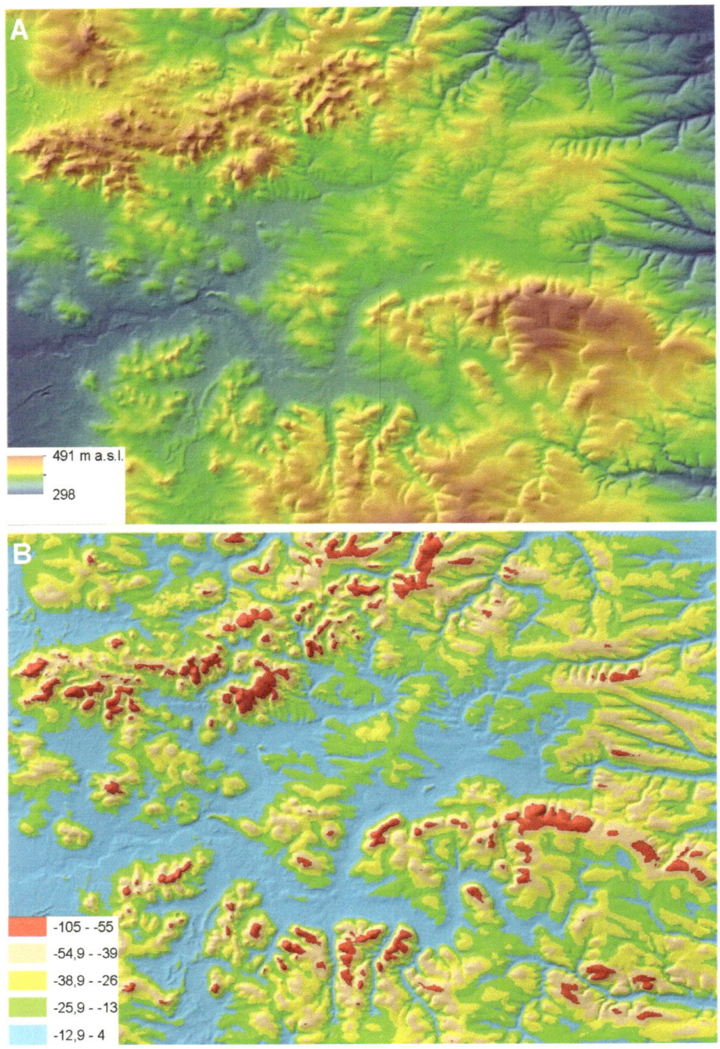

Figure 2. Elevation map (A), and simplified Hammond's landform classification (B).

nine possible hillslope units (**Figure 3**). They deliver a disjunctive description of the hillslope surface into units of homogeneous curvature characteristics. Then Dikau [124] proposed digital geomorphological relief model (DGRM) which will generate form facets and elements (as basic relief units) for geomorphological mapping and simulations for the derivation of more complex relief units. Dikau et al. [125] developed and applied an automated method for classifying macro landform types from DEM that was based on analysis of variation in topographic measures within areas defined by moving windows.

The original classification of Pennock et al. [126] explicitly assumed that surface form, as described by curvature, could be directly related to surface processes and to relative landform position (divergent/convergent shoulder, backslope footslope and level). Thus, strong profile convexity was assumed to be indicative of upper, water-shedding slope positions; whereas

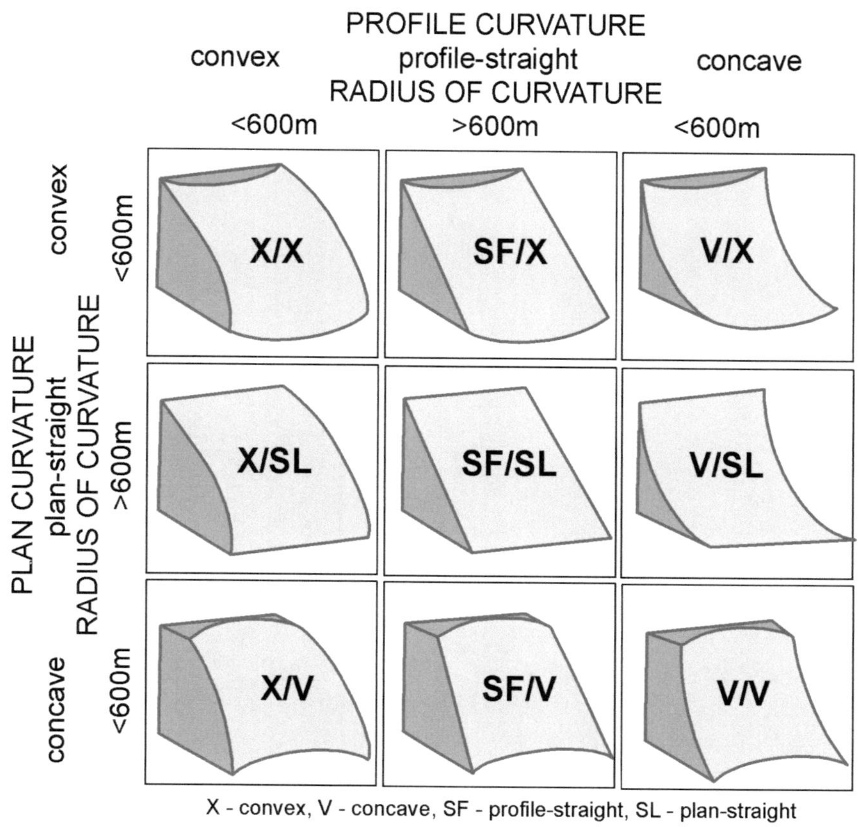

Figure 3. Classification of form elements by plan and profile curvature (after Dikau [124]).

strong profile concavity was associated with lower, water-receiving landform positions, and planar surfaces were associated with backslopes or flat areas. Of course, this pattern is not always adhered to and there are many instances where convex-concave patterns repeat over short distances along a longer hill slope [121].

Speight [121] described 10 morphological types of topographic landform positions: (i) crest, (ii) depression (open, closed), (iii) flat, (iv) slope, (v) simple slope, (vi) upper slope, (vii) mid-slope, (viii) lower slope, (ix) hillock, and (x) ridge (after [120]).

Very simple and interesting method for classifying relief on the base DEM is topographic position index (TPI). TPI is the difference between the elevation at a cell and the average elevation in a neighbourhood surrounding that cell (**Figure 4**). Positive values means that the cell is higher than its neighbours (indicate ridges, hills, etc.) while negative values means the cell is lower (indicate canyons, valleys, etc). TPI is a simplification of the landscape position index (LPI) described by Fels and Zobel [127] and was developed in detail by Weiss [128]. TPI values provide a simple and powerful means to classify the landscape into morphological classes [129]. TPI is naturally very scale-dependent, it means neighbourhood size (and shape) and DEM resolutions are critical to the final analysis, so the work with this index should be based on experiments with different threshold values.

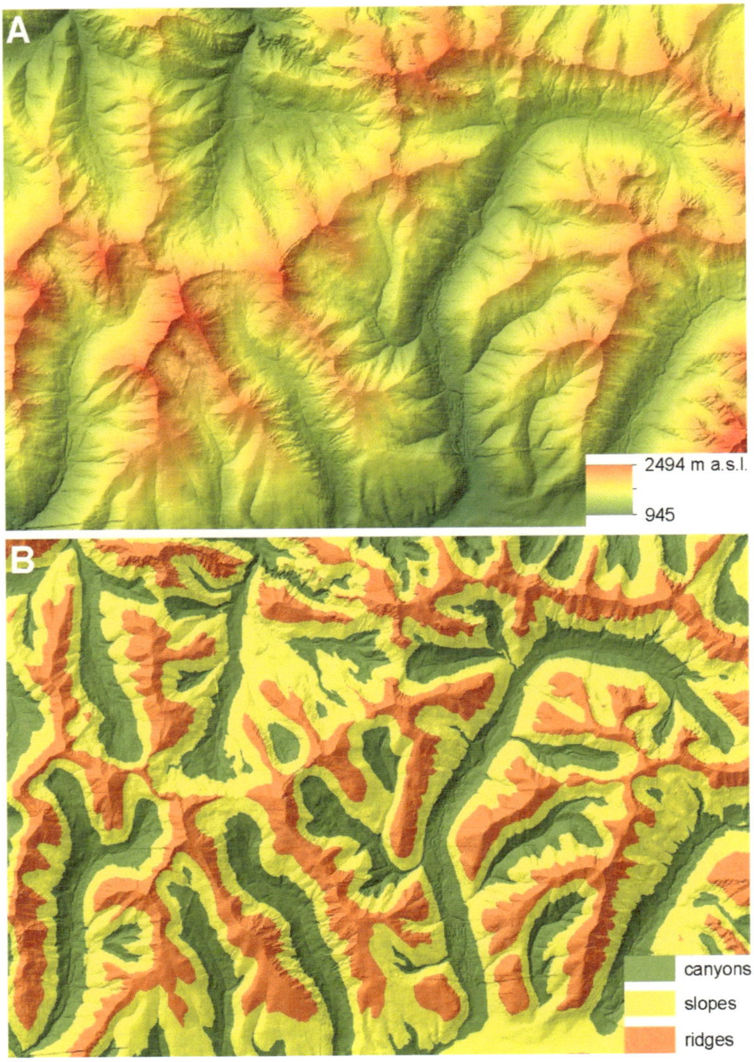

Figure 4. Elevation map (A), and TPI 3-category slope classification (B).

Drăguţ and Blaschke [130] presented an automated classification system of landform elements based on object-oriented image analysis. Firstly, one need to derive elevation, profile curvature, plan curvature, and slope gradient from DEM. Next, relatively homogenous objects are determined at several levels through segmentation of the image. These object primitives are classified as landform elements using a relative classification model, built both on the surface shape and on the altitudinal position of objects. The classification has nine classes which are peaks and toe slopes, steep and flat/gentle slopes, shoulders and negative contacts, head slopes, and side and nose slopes. Classes are defined using flexible fuzzy membership functions.

Iwahashi and Pike [88] developed an iterative procedure that classifies topography automatically into terrain types, grid cell by grid cell, on the basis of three morphometric variables, such as slope gradient, local convexity, and surface texture. They applied an unsupervised nested-means algorithm for 16 topographic types of the world. They noted the procedure is unsupervised and reflects frequency distributions of the input variables but not defined criteria. It causes that resulting classes are undefined and have to be calibrated empirically by subsequent analysis.

Within the context of defining landform units that maximize internal homogeneity and external differences, Minár and Evans [131] presented the concept of elementary forms (segments and units) defined by constant values of fundamental morphometric properties and limited by discontinuities of the properties. The basic system of form-defining properties represents altitude and its derivatives, constant values of which provide elementary forms with various types of homogeneity.

Drăguţ et al. [132] presented an algorithm to derive elementary forms from DEMs. Elementary forms were defined by constant values of fundamental morphometric properties and limited by discontinuities of these properties. A multi-resolution segmentation technique was customized to partition the layers of altitude derivatives into homogeneous divisions through a self-scalable procedure, which reveals the pattern encoded within the data. Layers were segmented successively, following the order of elevation derivatives (i.e. gradient, aspect, profile curvature, and plan curvature).

Jasiewicz and Stepinski [133] introduced a novel method for unsupervised classification and mapping of landforms from a DEM. This method involves the pattern recognition, not the differential geometry. The foundation of this idea is the concept of geomorphon (geomorphologic phonotypes), which is a simple ternary pattern that serves as an archetype of a particular terrain morphology. A finite number of 498 geomorphons constitute a comprehensive and exhaustive set of all possible morphological terrain types including standard elements of landscape as well as unfamiliar forms rarely found in natural terrestrial surfaces.

And one should remember that horizontal and vertical resolution of the elevation data used to present a terrain surface has a significant influence on the level of detail and the accuracy of portrayal of surface features and on the values of land-surface parameters that are computed from a DEM. One should first test out predictive efficiency for various DEM resolutions and neighbourhood sizes, and then objectively derive the most suitable resolution and search size [120].

Acknowledgements

The publication has been (partially) financed from the funds of the Leading National Research Centre (KNOW) received by the Centre for Polar Studies of the University of Silesia, Poland.

Author details

Bartłomiej Szypuła

Address all correspondence to: bartlomiej.szypula@us.edu.pl

University of Silesia in Katowice, Faculty of Earth Sciences, Sosnowiec, Poland

References

[1] de Margerie E. Geologie. Polybiblion Revue Bibliographique Universelle. Partie litteraire. 1886;**2**(24):310-330

[2] Bauer BO. Geomorphology. In: Goudie AS, editor. Encyclopedia of Geomorphology. London & New York: Taylor & Francis e-Library; 2006. pp. 428-435

[3] Huggett RJ. Fundamentals of Geomorphology. Routledge Fundamentals of Physical Geography. New York: Taylor & Francis e-Library; 2005. p. 386

[4] Klimaszewski M. Geomorfologia. 6th ed. Warszawa: PWN; 1980. p. 1063

[5] Davis WM. The rivers and valleys of Pennsylvania. National Geographical Magazine. 1889;**1**: 183-253. (Also in Geographical Essays)

[6] Davis WM. The geographical cycle. Geographical Journal. 1899;**14**:481-504

[7] Davis WM. Geographical Essays. Boston, Mass.: Ginn; 1909

[8] Penck W. Die morphologische Analyse, ein Kapitel der physikalischen Geologie. Stuttgart: Engelhorn; 1924

[9] Penck W. Morphological Analysis of Landforms. London: Macmillan; 1953

[10] Migoń P. Geomorfologia. 1st ed. Warszawa: PWN; 2006. p. 461

[11] Horton RE. Drainage basin characteristics. Transactions, American Geophysical Union. 1932;**14**:350-361

[12] Horton RE. Erosional development of streams and their drainage basins; hydrophysical approach to quantitative morphology. Bulletin of the Geological Society of America. 1945;**56**:275-370

[13] Strahler AN. Quantitative analysis of watershed geomorphology. Transactions, American Geophysical Union. 1957;**38**:913920

[14] Strahler AN. Hypsometric (area-altitude) analysis of erosional topography. Geological Society of America Bulletin. 1952;**63**:1117-1142. DOI:10.1130/0016-7606(1952)63[1117:HAAOET]2.0.CO;2

[15] Strahler AN. Quantitative slope analysis. Geological Society of America Bulletin. 1956;**67**: 571-596

[16] Strahler AN. Quantitative geomorphology of drainage basins and channel networks. In: Chow VT, editor. Handbook of Applied Hydrology. New York: McGraw-Hill; 1964. pp. 439-476

[17] Schumm S. The relation of drainage basin relief to sediment loss. International Association of Hydrological Sciences Publications. 1954;**36**:216-219

[18] Chorley RJ. The drainage basin as the fundamental geomorphic unit. In: Chorley RJ, editor. Water, Earth and Man. London: Methuen; 1969. pp. 77-100

[19] Williams PW. Morphometric analysis of polygonal karst in New Guinea. Geological Society of America Bulletin. 1972;**83**:761-796

[20] Evans IS. The geomorphology and morphometry of glacial and nival areas. In: Chorley RJ, editor. Water, Earth and Man. London: Methuen; 1969. pp. 369-380

[21] Evans IS. The geomorphometry and asymmetry of glaciated mountains [Unpubl. Ph.D. Thesis]. University of Cambridge; 1974

[22] Lustig LK. Quantitative analysis of desert topography. Arid Lands in Perspective. 1969;45-48

[23] Evans IS. General geomorphometry, derivatives of altitude, and descriptive statistics. In: Chorley RJ, editor. Spatial Analysis in Geomorphology. Harper & Row, New York. 1972. pp. 17-90

[24] Pike RJ. Geomorphometry – progress, practice, and prospect. Zeitschrift fur Geomorphologie, Supplementband. 1995;**101**:221-238

[25] Pike RJ. A Bibliography of Terrain Modeling (Geomorphometry), the Quantitative Representation of Topography - Supplement 4.0. (USGS, Open File Report 02-465). Menlo Park, Calif.: US Department of the Interior, United States Geological Survey; 2002

[26] Evans IS. Geomorphometry. In: Goudie AS, editor. Encyclopedia of Geomorphology. London & New York: Taylor & Francis e-Library; 2006. pp. 435-439

[27] Pike RJ. Digital elevation model. In: Goudie AS, editor. Encyclopedia of Geomorphology. London & New York: Taylor & Francis e-Library; 2006. pp. 260-263

[28] Meyer W. Concepts of Mathematical Modelling. New York: McGraw-Hill Book Company; 1995

[29] Miller CL, Laflamme RA. The digital terrain model — Theory and application. Photogrammetric Engineering. 1958;**24**(3):433-442

[30] Burrough PA. Methods of interpolation. In: Principles of Geographical Information Systems for Land Resources Assessment. Oxford: OUP; 1986. pp. 147-166

[31] Moore ID, Grayson RB, Ladson AR. Digital terrain modelling: A review of hydrological, geomorphological, and biological applications. Hydrological Processes. 1991;**5**(1):3-30. DOI: 10.1002/hyp.3360050103

[32] Moore ID, Turner AK, Wilson JP, Jenson S, Band L. GIS and land-surface–subsurface process modelling. In: Goodchild MF, Parkes BO, Steyaert LT, editors. Environmental Modeling with GIS. New York: Oxford University Press; 1993. pp. 196-230

[33] Weibel R., Heller M. Digital terrain modeling. In: Maguire D, Goodchild M, Rhind D, editors. Geographical Information Systems. London: Longman Scientific and Technical; 1991. pp. 269-297

[34] Li Z, Zhu Q, Gold C. Digital terrain modeling: Principles and methodology. Boca Raton: CRC Press; 2005. p. 319

[35] Hengl T, Evans IS. Mathematical and digital models of the land surface. In: Hengl T, Reuter HI, editors. Geomorphometry. Concepts, Software, Applications. Elsevier; 2009. pp. 31-64. DOI: 10.1016/S0166-2481(08)00002-0

[36] Wilson JP, Gallant JC. Digital terrain analysis. In: Wilson JP, Gallant JC, editors. Terrain Analysis: Principles and Applications. John Wiley & Sons, Inc.; 2000. pp. 1-28

[37] Wise SM. The effect of GIS interpolation errors on the use of digital elevation models in geomorphology. In: Lane SN, Richards KS, Chandler JH, editors. Landform Monitoring, Modelling and Analysis. Wiley; 1998. pp. 139-164

[38] Zevenbergen LW, Thorne CR. Quantitative analysis of land surface topography. Earth Surface Processes and Landforms. 1987;**12**:47-56

[39] Carter JR. Digital representations of topographic surfaces. Photogrammetric Engineering and Remote Sensing. 1988;**54**(11):1577-1580

[40] Hormann K. Geomorphologische Kartenanalyse mit Hilfe elektronischer Rechenanlagen. Zeitschrift für Geomorphologie. 1969;**133**(1):75-98

[41] Moore ID, Grayson RB. Terrain-based catchment partitioning and runoff prediction using vector elevation data. Water Resources Research. 1991;**27**:1177-1191

[42] Wood J. Digital elevation model (DEM). In: Kemp KK, editor. Encyclopedia of Geographic Information Science. SAGE Publications, Inc; 2008. Los Angeles. pp. 107-109. DOI: 10.4135/9781412953962

[43] Nelson A, Reuter HI, Gessler P. DEM production methods and sources. In: Hengl T, Reuter HI, editors. Geomorphometry. Concepts, Software, Applications. Elsevier, Amsterdam; 2009. pp. 65-86. DOI: 10.1016/S0166-2481(08)00003-2

[44] Baltsavias EP. A comparison between photogrammetry and laser scanning. ISPRS Journal of Photogrammetry and Remote Sensing, 1999;**54**:83-94

[45] Smith SE. Topographic mapping. In: Grunwald S, editor. Environmental Soil–Landscape Modeling: Geographic Information Technologies and Pedometrics. Vol. 1. New York: CRC Press; 2005. pp. 155-182

[46] Gesch DB, Greenlee S. GTOPO30 Documentation. U.S. Geological Survey; 1997. p. 19

[47] DMA (Defense Mapping Agency). Defense Mapping Agency Product Specifications for Digital Terrain Elevation Data (DTED). 2nd ed. St. Louis, Missouri: Defense Mapping Agency Aerospace Center; 1986. p. 26

[48] DMA (Defense Mapping Agency). Digitizing the Future. 3d ed. Washington, D.C.: Defense Mapping Agency; 1990. p. 105

[49] USGS (United States Geological Survey). Digital Elevation Models, data User Guide 5. Reston, Virginia, USGS; 1993. p. 50

[50] Hastings DA, Dunbar PK. Global Land One-kilometer Base Elevation (GLOBE) Digital Elevation Model, Documentation, Volume 1.0. Key to Geophysical Records

Documentation (KGRD) 34. 325 Broadway, Boulder, Colorado 80303, U.S.A: National Oceanic and Atmospheric Administration, National Geophysical Data Center; 1999

[51] Hastings DA. Global Land One-km Base Elevation (GLOBE) DEM. Southern Africa Subset; 2000. p. 6

[52] Rodriguez E, et al. An assessment of the SRTM topographic products [Technical Report JPL D-31639]. Pasadena, California: Jet Propulsion Laboratory; 2005. p. 143

[53] Farr TG, et al. The shuttle radar topography mission. Journal of Geophysical Research. 2007;**45**:43. RG2004, DOI: 10.1029/2005RG000183

[54] Amante C, Eakins BW. ETOPO1 1 arc-minute global relief model: Procedures, data sources and analysis. Boulder, Colorado: National Geophysical Data Center Marine Geology and Geophysics Division; 2009. p. 19

[55] NASA JPL. ASTER Global Digital Elevation Model [Data set]. NASA JPL; 2009. DOI: 10.5067/ASTER/ASTGTM.002

[56] Tachikawa T, Hato M, Kaku M, Iwasaki A. The characteristics of ASTER GDEM version 2; IGARSS, July 2011

[57] Danielson JJ, Gesch DB. Global multi-resolution terrain elevation data 2010 (GMTED2010): U.S. Geological Survey Open-File Report 2011-1073. 2011. p. 26

[58] Tadono T, Ishida H, Oda F, Naito S, Minakawa K, Iwamoto H. Precise global DEM generation by ALOS PRISM. ISPRS Annals of the Photogrammetry, Remote Sensing and Spatial Information Sciences. 2014;**II-4**:71-76. DOI: 10.5194/isprsannals-II-4-71-2014

[59] Takaku J, Tadono T, Tsutsui K. Generation of high resolution Global DSM from ALOS PRISM. The International Archives of the Photogrammetry, Remote Sensing and Spatial Information Sciences, Vol. XL-4, ISPRS TC IV Symposium, Suzhou, China; 2014. p. 243-248. DOI: 10.5194/isprsarchives-XL-4-243-2014.

[60] Mark DM. Geomorphometric parameters: a review and evaluation. Geografiska Annaler. 1975;**57A**(3-4):165-177

[61] Montgomery DR. Relief. In: Goudie AS, editor. Encyclopedia of Geomorphology. London & New York: Taylor & Francis e-Library; 2006. pp. 840-842

[62] Deng Y, Wilson JP, Gallant JC. Terrain analysis. In: Wilson JP, Fotheringham AS, editors. The Handbook of Geographic Information Science. 2008. pp. 417-435. DOI: 10.1002/9780470690819.ch23

[63] Chorley RJ. Climate and morphometry. Journal of Geology. 1957;**65**:628-638

[64] Chorley RJ. Group operator variance in morphometric work with maps. American Journal of Science. 1958;**256**:208-218

[65] Chorley RJ, Morgan MA. Comparison of morphometric features, Unaka Mountains, Tennessee and North Carolina, and Dartmoor, England. Bulletin, Geological Society of America. 1962;**73**:17-34

[66] Stone RO, Dugundji J. A Study of Microrelief: Its Mapping, Classification, and Quantification by Means of a Fourier Analysis. Amsterdam: Elsevier; 1965. p. 97

[67] Hobson RD. Fortran IV programs to determine surface roughness in topography for the CDC 3400 computer. Kansas Geological Survey Computer Contribution. 1967;**14**:28

[68] Wood W.F, Snell JB. A Quantitative System for Classifying Landforms [Technical Report EP-124]. Natick, MA: U.S. Army Quartermaster Research and Engineering Center; 1960. p. 20

[69] Krcho J. Morphometric analysis of relief on the basis of geometric aspect of field theory. Acta geographica Universitatis Comenianae, Geographico-physica. 1973;**1**:11-233

[70] Moore RF. Thornes JB. Leap - a suite of Fortran IV programs for generating erosional potentials of land surfaces from topographic information. Computers & Geosciences. 1976;**2**:493-499

[71] Pike RJ. Information content of planetary terrain: varied effectiveness of parameters for the Earth. In: Lunar and Planetary Science Conference. vol. 18; 1987

[72] Pike RJ. The geometric signature: quantifying landslide-terrain types from digital elevation models: Mathematical Geology. 1988;**20**:491-512. DOI: 10.1007/BF00890333

[73] Pike RJ. Geometric signatures-experimental design, first results. In: Ohmori H, editor. DEMs and Geomorphometry, Special Publications of the Geographic Information Systems Association, Proceedings of the Symposia on New Concepts and Modeling in Geomorphology and Geomorphometry, DEMs and GIS; In: Fifth International Conference on Geomorphology; 24-26 august 2001; Tokyo. pp. 50-51

[74] Pike RJ, Acevedo W, Card DH. Topographic grain automated from digital elevation models. In: Proceedings of the Ninth International Symposium on Computer Assisted Cartogtraphy. Baltimore, MD: ASPRS/ASCM; 1989. pp. 128-137

[75] Speight JG. A parametric approach to landform regions. Special Publication Institute of British Geographers. 1974;**7**:213-230

[76] Speight JG. The role of topography in controlling throughflow generation: A discussion. Earth Surface Processes and Landforms. 1980;**5**:187-191

[77] Schmidt J, Merz B, Dikau R. Morphological structure and hydrological process modelling. Zeitschrift für Geomorphologie. Suppl.-Bd. 1998;**112**:55-66

[78] Schmidt J, Dikau R. Extracting geomorphometric attributes and objects from digital elevation models — semantics, methods, future needs. In: Dikau R, Saurer H, editors. GIS for Earth Surface Systems — Analysis and Modelling of the Natural Environment. Schweizbart'sche Verlagsbuchhandlung; 1999. pp. 153-173

[79] Basso B. Digital terrain analysis: Data source, resolution and applications for modeling physical processes in agroecosystems. Rivista Italiana di Agrometeorologia. 2005;**2**:5-14

[80] Wilson JP, Bishop MP. Geomorphometry. In: Shroder JF, Bishop MP, editors. Treatise on Geomorphology, vol 3, Remote Sensing and GIScience in Geomorphology. San Diego: Academic Press; 2013. pp. 162-186

[81] Bork HR, Rohdenburg H. Transferable parameterization methods for distributed hydrological and agroecological catchment models. Catena. 1986;**13**:99-117

[82] Shary PA, Sharaya SL, Mitusov AV. Fundamental quantitative methods of land surface analysis. Geoderma. 2002;**107**:1-32. DOI: 10.1016/S0016-7061(01)00136-7

[83] Pennock DJ. Terrain attributes, landform segmentation, and soil redistribution. Soil and Tillage Research, 2003;**69**(1-2):15-26.

[84] Goodwin CN, Tarboton DG. Morphometric properties. In: Goudie AS, editor. Encyclopedia of Geomorphology. London & New York: Taylor & Francis e-Library; 2006. pp. 696-698

[85] Hadley RF, Schumm SA. Sediment Sources and Drainage-Basin Characteristics in the Upper Cheyenne River Basin, Washington. US Geological Survey Water Supply Paper; 1961. p. 1,531

[86] Olaya V. Basic Land-Surface parameters. In: Hengl T, Reuter HI, editors. Geomorphometry. Concepts, Software, Applications. Elsevier, Amsterdam; 2009. pp. 141-169. DOI: 10.1016/S0166-2481(08)00006-8

[87] Evans IS, Minár J. A classification of geomorphometric variables. In: International Geomorphometry 2011; May 30 2011; Redlands, CA: Geomorphometry.org;. 2011. p. 105-108

[88] Iwahashi J, Pike RJ. Automated classifications of topography from DEMs by an unsupervised nested-means algorithm and a three-part geometric signature. Geomorphology. 2007;**86**(3-4):409-440. DOI: 10.1016/j.geomorph.2006.09.012

[89] Florinsky I.V.: Accuracy of local topographic variables derived from digital elevation models. International Journal of Geographical Information Science. 1998;**12**(1):47-62.

[90] Riley SJ, DeGloria SD, Elliot R. A terrain ruggedness index that quantifies topographic heterogeneity. Intermountain Journal of Sciences. 1999;**5**:23-27

[91] Pike RI, Wilson SE. Elevation-relief ratio, hypsometric integral and geomorphic area altitude analysis. Geological Society of America Bulletin. 1971;**82**:1079-1084

[92] Yokoyama R, Sirasawa M, Pike RJ. Visualizing topography by openness: A new application of image processing to digital elevation models. Photogrammetric Engineering & Remote Sensing. 2002;**68**(3):257-265

[93] McNab WH. Terrain shape index: quantifying effect of minor landforms on tree height. Forest Science. 1989;**35**(1):91-104

[94] McNab WH. A topographic index to quantify the effect of mesascale landform on site productivity. Canadian Journal of Forest Research. 1992;**23**:1100-1107

[95] McCune B, Keon D. Equations for potential annual direct incident radiation and heat load index. Journal of Vegetation Science. 2002;**13**:603-606

[96] Bishop TFA, Minasny B. Digital soil-terrain modelling: The predictive potential and uncertainty. In: Grunwald S, editor. Environmental Soil–Landscape Modeling: Geographic Information Technologies and Pedometrics. Boca Raton, FL: CRC Press; 2005. pp. 185-213

[97] Hengl T, Gruber S, Shrestha DP. Digital Terrain Analysis in ILWIS. Lecture Notes. Enschede: International Institute for Geo-Information Science & Earth Observation (ITC); 2003. p. 56

[98] Schumm SA. Evolution of drainage systems and slopes in badlands at Perth Amboy, New Jersey. Geological Society of America Bulletin. 1956;**67**(5):597-646. DOI: 10.1130/0016-7606(1956)67[597:EODSAS]2.0.CO;2

[99] Melton MA. An analysis of the relation among elements of climate, surface properties and Geomorphology [ONR Technical Report 11]. New York: Columbia University; 1957

[100] Melton MA. Correlation structure of morphometric properties of drainage systems and their controlling agents. Journal of Geology. 1958;**66**:442-460

[101] Engstrom WN. Morphometric analysis of mountain drainage basins in the Basin and Range Province, USA. Zeitschrift für Geomorphologie N. F. 1989;**33**:443-453

[102] Cannon PJ. Generation of explicite parameter for a quantitative geomorphic study of the Mill Creek drainage basin. Oklahoma Geology Notes. 1976;**36**(1):3-16

[103] Ramírez-Herrera MT. Geomorphic assessment of active tectonics in the Acambay graben, Mexican Volcanic Belt. Earth Surface Processess and Landforms. 1998;**23**:317-332

[104] Jenness J. Surface Areas and Ratios from Elevation Grid (surfgrids.avx) extension for ArcView 3.x, v. 1.2 Jenness Enterprises [Internet]. 2002. Available from: http://www.jennessent.com/arcview/surface_areas.htm (Accessed on: 1 November 2011)

[105] Bull WB, McFadden LD. Tectonic geomorphology north and south of the Garlock Fault, California. In: Doehring DO, editor. Geomorphology in Arid Regions: Annual Binghamton Conference. State University of New York at Binghamton; 1977. pp. 115-136

[106] Sharma HS, editor. Perspectives in Geomorphology. Essays on Indian Geomorphology. vol. 4. New Delhi: Concept Publishing; 1982. p. 358

[107] Szypuła B. Relief Index (RI) as a simple tool for geomorphometry. In: Jasiewicz J, Zwoliński Z, Mitasova H, Hengl T, editors. Geomorphometry for Geosciences. Poznań: Adam Mickiewicz University in Poznań - Institute of Geoecology and Geoinformation, International Society for Geomorphometry; 2015. pp. 127-128

[108] SAGA-GIS Tool Library Documentation v3.0.0 [Internet]. 2016. Available from: http://www.saga-gis.org/en/index.htm (Accessed on: 21 December 2016)

[109] Conrad O, Bechtel B, Bock M, Dietrich H, Fischer E, Gerlitz L, Wehberg J, Wichmann V, Böhner J. System for Automated Geoscientific Analyses (SAGA) v. 2.1.4. Geoscientific Model Development. 2015;**8**:1991-2007. DOI: 10.5194/gmd-8-1991-2015

[110] ESRI (Environmental Systems Resource Institute). ArcGIS Desktop 10.5. Redlands, California; 2016

[111] Evans IS, Hengl T, Gorsevski P. Applications in geomorphology. In: Hengl T, Reuter HI, editors. Geomorphometry. Elsevier, Amsterdam; 2009. pp. 497-526. DOI: 10.1016/S0166-2481(08)00022-6

[112] Hammond EH. Small-scale continental landform maps. Annals of the Association of American Geographers. 1954;**44**(1):33-42

[113] Anstey RL. Physical characteristics of alluvial fans. Technical Report ES-20, US Army Natick Labs; 1965

[114] Anstey RL. Application of Quantified Terrain Data to Vehicle Design Criteria. In: Proceedings of the 5th US Army Science Conference. West Point; 1966

[115] Hammond EH. Analysis of properties in land form geography: An application to broad-scale land form mapping. Annals of the Association of American Geographers. 1964;**54**:11-19

[116] Peucker TK, Douglas DH. Detection of Surface-Specific points by local parallel processing of discrete terrain elevation data. Computer Graphics and Image Processing. 1975;**4**(4):375-387. DOI: 10.1016/0146-664X(75)90005-2

[117] Ruhe RV. Elements of the soil landscape. In: Transactions of the 9th Congress of the International Society of Soil Science. Vol. 4. Madison, WI: International Soil Science Society; 1960. pp. 165-170

[118] Dalrymple JB, Blong RJ, Conacher AJ. A hypothetical nine unit landsurface model. Zeitschrift für Geomorphologie. 1968;**12**:60-76

[119] Conacher AJ, Dalrymple JB. The nine-unit landsurface model: an approach to pedogeomorphic research. Geoderma. 1977;**18**(1-2):1-154

[120] MacMillan RA, Shary PA. Landforms and landform elements in geomorphometry. In: Hengl T, Reuter HI, editors. Geomorphometry. Concepts, Software, Applications. Elsevier, Amsterdam; 2009. pp. 227-254. DOI: 10.1016/S0166-2481(08)00009-3

[121] Speight JG. Landform. In: McDonald RC, Isbell RF, Speight JG, Walker J, editors. Australian Soil and Land Survey: Field Handbook. Melbourne: Inkata Press; 1990. pp. 9-57

[122] Dikau R, Rasemann S, Schmidt J. Hillslope, Form. In: Goudie AS, editor. Encyclopedia of Geomorphology. London & New York: Taylor & Francis e-Library; 2006. pp. 516-521

[123] Young A. Slopes. London: Longman; 1972. p. 268

[124] Dikau R. The application of a digital relief model to landform analysis in geomorphology. In: Raper J, editor. Three-dimensional Applications in Geographical Information Systems. London: Taylor and Francis; 1989. pp. 51-77

[125] Dikau R, Brabb EE, Mark RM. Landform classification of New Mexico by computer. Open File Report 91-634. U.S. Geological Survey; 1991. p. 15

[126] Pennock DJ, Zebarth BJ, De Jong E. Landform classification and soil distribution in hummocky terrain, Saskatchewan, Canada. Geoderma. 1987;**40**(3-4):297-315. DOI: 10.1016/0016-7061(87)90040-1

[127] Fels JE, Zobel R. Landscape position and classified landtype mapping for statewide DRASTIC mapping project. North Carolina State University Technical Report VEL. 95.1; 1995

[128] Weiss A. Topographic positions and landforms analysis (Conference Poster). ESRI International User Conference; 2001; San Diego, CA. p. 9-13

[129] Jenness J. Topographic position index (tpi_jen.avx) extension for ArcView 3.x. Jenness Enterprises [Internet]. 2005. Available from: http://www.jennessent.com (Accessed: 11 December 2017)

[130] Drăguţ L, Blaschke T. Automated classification of landform elements using object-based image analysis. Geomorphology. 2006;**81**:330-344. DOI: 10.1016/j.geomorph.2006.04.013

[131] Minár J, Evans IS. Elementary forms for land surface segmentation: The theoretical basis of terrain analysis and geomorphological mapping. Geomorphology. 2008;**95**(3-4):236-259. DOI: 10.1016/j.geomorph.2007.06.003

[132] Drăguţ L, Csillik O, Minár J, Evans IS. Land-surface segmentation to delineate elementary forms from Digital Elevation Models. In: Geomorphometry 2013; 16-20 october 2013; Nanjing, China; 2013

[133] Jasiewicz J, Stepinski TF. Geomorphons — A pattern recognition approach to classification and mapping of landforms. Geomorphology. 2013;**182**(2013):147-156. DOI: 10.1016/j.geomorph.2012.11.005